Cambridge Elements ≡

Elements in Grid Energy Storage
edited by
Babu Chalamala
Sandia National Laboratories
Vincent Sprenkle
Pacific Northwest National Laboratory
Imre Gyuk
US Department of Energy
Ralph D. Masiello
Quanta Technology
Raymond Byrne
Sandia National Laboratories

ENERGY STORAGE APPLICATIONS IN TRANSMISSION AND DISTRIBUTION GRIDS

Hisham Othman
Quanta Technology LLC

CAMBRIDGE
UNIVERSITY PRESS

University Printing House, Cambridge CB2 8BS, United Kingdom

One Liberty Plaza, 20th Floor, New York, NY 10006, USA

477 Williamstown Road, Port Melbourne, VIC 3207, Australia

314–321, 3rd Floor, Plot 3, Splendor Forum, Jasola District Centre,
New Delhi – 110025, India

103 Penang Road, #05–06/07, Visioncrest Commercial, Singapore 238467

Cambridge University Press is part of the University of Cambridge.

It furthers the University's mission by disseminating knowledge in the pursuit of education, learning, and research at the highest international levels of excellence.

www.cambridge.org
Information on this title: www.cambridge.org/9781009014038
DOI: 10.1017/9781009029223

© Cambridge University Press & Assessment 2022

First published 2022

A catalogue record for this publication is available from the British Library.

ISBN 978-1-009-01403-8 Paperback
ISSN 2634-9922 (online)
ISSN 2634-9914 (print)

Energy Storage Applications in Transmission and Distribution Grids

Elements in Grid Energy Storage

DOI: 10.1017/9781009029223
First published online: June 2022

Hisham Othman
Quanta Technology LLC

Author for correspondence: Hisham Othman,
hothman@Quanta-technology.com

Abstract: The application of energy storage within transmission and distribution grids as non-wire alternative solutions (NWS) is hindered by the lack of readily available analysis tools, standardized planning processes, and practical know-how. This Element provides a theoretical basis along with examples and real-world case studies to guide grid planners in the siting, sizing, and lifetime techno-economic evaluation of storage systems. Many applications are illustrated, including feeder peak shaving, feeder reliability improvements, transmission reliability, transmission congestion relief, and renewable integration. Three case studies, based on the author's consulting experience, illustrate the versatility of the analysis methods and provide a guide to grid planners while tackling real-world problems.

Keywords: non-wire solutions, T&D applications, congestion relief, feeder reliability, energy storage

ISBNs: 9781009014038 (PB), 9781009029223 (OC)
ISSNs: 2634-9922 (online), 2634-9914 (print)

Contents

1 Introduction

Energy storage is increasingly attracting the interest of policy makers, regulators, project developers, and electric utilities. These modular assets can rapidly inject and absorb not only reactive power but also real power, making them of universal utility along the electricity value chain from generation plants to consumers. U.S. Department of Energy and Sandia National Labs [1] [2] published a comprehensive list of 17 applications that serve multiple stakeholders: behind-the-meter (BTM), distribution, transmission, generation, and wholesale markets, as shown in Figure 1-1. With the decreasing chemistry prices and deepening of the supply chain, energy storage applications are increasingly economical as compared to conventional solutions. The focus of this section is on Transmission and Distribution (T&D) applications.

Electricity flows on T&D lines according to Kirchhoff's laws, based on the levels and locations of power injections and withdrawals, the impedances of the lines, and the topology of the network. To maintain reliability, power injections and withdrawals are constrained to avoid violating one or more of the system operating criteria in the form of line current overloads, voltage violations, dynamic or transient instabilities, flicker, harmonic distortions, or subsynchronous resonances during normal system operating conditions or after the onset of a preselected set of planning contingencies. An obvious example is the case of limiting the combined flow on two parallel transmission lines to the lower emergency rating of either line to avoid overloads should one of the lines experience a sudden outage. This operating philosophy indeed causes the average loading of transmission lines as a percentage of their rating, in most systems, to be less than 50% at any time. These limitations on power injections and withdrawals, although prudent and necessary, have enormous economic consequences. Quite often, they are triggered by loading patterns that are infrequent and may last only a few hours in a day or in a season.

Energy storage can inject and withdraw electric power within its design and operating limits, and thus, when located appropriately, can help expand the flow capability of the grid by absorbing the excess power or supplying the deficit power during those constrained hours, either continuously or after the onset of a contingency. One should be aware that if the level or persistence of overloads is high, energy storage will not be a viable technical or economical solution when compared to other traditional solutions.

The rapid dynamic response of a storage system while injecting and absorbing active power and, when equipped with a four-quadrant inverter, reactive power, and its readily available reservoir of energy, albeit limited, can help system planners to increase the transfer limits on congested interfaces that are

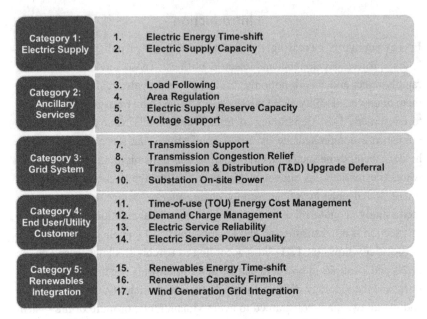

Category 1: **Electric Supply**	1.	**Electric Energy Time-shift**
	2.	**Electric Supply Capacity**
Category 2: **Ancillary** **Services**	3.	**Load Following**
	4.	**Area Regulation**
	5.	**Electric Supply Reserve Capacity**
	6.	**Voltage Support**
Category 3: **Grid System**	7.	**Transmission Support**
	8.	**Transmission Congestion Relief**
	9.	**Transmission & Distribution (T&D) Upgrade Deferral**
	10.	**Substation On-site Power**
Category 4: **End User/Utility** **Customer**	11.	**Time-of-use (TOU) Energy Cost Management**
	12.	**Demand Charge Management**
	13.	**Electric Service Reliability**
	14.	**Electric Service Power Quality**
Category 5: **Renewables** **Integration**	15.	**Renewables Energy Time-shift**
	16.	**Renewables Capacity Firming**
	17.	**Wind Generation Grid Integration**

Figure 1.1 Energy storage can provide up to 17 services

constrained post contingencies either by thermal limits, stability limits, or voltage limits. It can also counteract the effects of solar intermittency in distribution networks and thus increase the hosting capacity of distributed energy resources (DERs) and provide differentiated delivery reliability or resilience wherever required. A properly sized and sited storage system can provide several grid-reliability and efficiency services while facilitating public policy initiatives, as summarized in Figure 1–2.

The objective of this Element is to present methods of analysis to help transmission and distribution planners to properly examine the technical and economical efficacy of energy storage as a non-wire solution (NWS) to address not only grid reliability but also grid performance. Key emphasis will be on siting, sizing, revenue stacking, and techno-economic lifetime analysis. The scope of this Element excludes other important planning aspects such as control system design including selection of measurement signals and controller parameter tuning, and protection system design and relay settings.

1.1 Strategic Value of Energy Storage as a T&D Resource

Electric power systems are inherently complex to manage. The instantaneous production, transport, and delivery of energy requires multiple layers of autonomous protection and control systems, primarily at the generation plants, T&D

grid, load, and supervisory control centers, to ensure the quality of delivery in this massive just-in-time power system. Energy storage can be right-sized and managed to add strategic value to existing systems as follows:

- Improve the utilization of existing T&D lines by time shifting power injections and withdrawals to reduce system overloads and in so doing increase the flow limits.
- Provide substantial flexibility to help position the grid gradually on a modernization path to economically accommodate new forms of intermittent energy sources and consumption patterns.
- Leverage the modularity of energy storage to de-risk grid investments especially in the face of growing planning uncertainties and lengthy permitting processes.

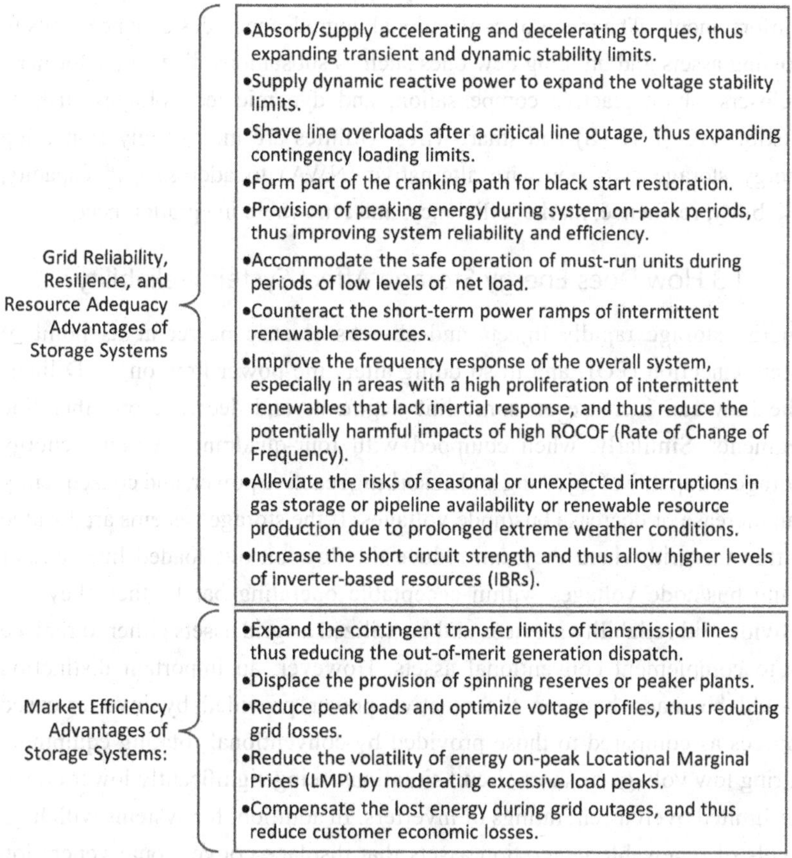

Grid Reliability, Resilience, and Resource Adequacy Advantages of Storage Systems	•Absorb/supply accelerating and decelerating torques, thus expanding transient and dynamic stability limits. •Supply dynamic reactive power to expand the voltage stability limits. •Shave line overloads after a critical line outage, thus expanding contingency loading limits. •Form part of the cranking path for black start restoration. •Provision of peaking energy during system on-peak periods, thus improving system reliability and efficiency. •Accommodate the safe operation of must-run units during periods of low levels of net load. •Counteract the short-term power ramps of intermittent renewable resources. •Improve the frequency response of the overall system, especially in areas with a high proliferation of intermittent renewables that lack inertial response, and thus reduce the potentially harmful impacts of high ROCOF (Rate of Change of Frequency). •Alleviate the risks of seasonal or unexpected interruptions in gas storage or pipeline availability or renewable resource production due to prolonged extreme weather conditions. •Increase the short circuit strength and thus allow higher levels of inverter-based resources (IBRs).
Market Efficiency Advantages of Storage Systems:	•Expand the contingent transfer limits of transmission lines thus reducing the out-of-merit generation dispatch. •Displace the provision of spinning reserves or peaker plants. •Reduce peak loads and optimize voltage profiles, thus reducing grid losses. •Reduce the volatility of energy on-peak Locational Marginal Prices (LMP) by moderating excessive load peaks. •Compensate the lost energy during grid outages, and thus reduce customer economic losses.

Figure 1.2 Advantages of storage systems for grid reliability and market efficiency

1.2 Role of Energy Storage in T&D Planning

Transmission and distribution (T&D) grids are operated under strict reliability standards that do not allow lines and transformers to overload beyond their thermal ratings and do not allow bus voltages to wander outside acceptable bounds. The reliability standards address operating criteria not only under normal or intact (i.e., N-0) conditions but also under potential contingencies (N-1, N-1–1) that take one or multiple grid elements out of service. The reliability standards ensure robustness and security of the extra-high voltage grid to avoid the risk of cascading outages and widespread loss of load. They also assure customers served from the distribution grid of an agreed-upon level of service quality.

Within a typical planning horizon that spans 5–20 years, system planners update the demand forecast and consider the impact of the planned retirements of generation assets, the interconnection queue of resources, and the health of existing T&D assets on grid reliability and the need for T&D upgrades and reinforcements. The menu of conventional upgrades includes enhancements to existing assets and building new ones such as substations, lines, transformers, reclosers, shunt reactive compensation, and dynamic technologies such as Flexible AC (FACTS) and smart wires. Utilities are increasingly evaluating energy storage as a non-wire alternative (NWA) to address grid capacity, reliability, resilience, market efficiency, and renewable integration needs.

1.3 How Does Energy Storage Affect System Reliability?

Energy storage rapidly injects and absorbs electric power at its point of interconnection (POI) and in so doing alters the power flow on T&D lines. The flow can increase on some line segments and decrease on other line segments. Similarly, when equipped with four-quadrant inverters, energy storage is capable of injecting and absorbing reactive power, and consequently can increase or decrease bus/node voltages. If the storage systems are located within the grid where they can reduce the flow on overloaded lines and/or bring bus/node voltages within acceptable operating bands, then they can provide grid reliability services and be utilized as grid assets either to replace or to complement conventional assets. However, an important distinction should be made between the reactive power provided by inverter-based sources as compared to those provided by conventional rotating equipment during low voltage or short circuit situations being significantly lower due to the limited overcurrent ratings of inverters. In addition, for systems with high levels of renewable generation assets that displace conventional generation assets, the same storage systems can additionally provide other essential reliability services including:

- Inertial frequency response to limit the rate of change of frequency (RoCoF) within mandated limits,
- Primary frequency response to control the frequency excursions without triggering load-shedding schemes,
- Stiffening the grid's short-circuit strength to enable proper operation of renewable inverters, and
- Mitigating potential flicker concerns by renewable inverters.

1.4 Should the Storage System Charge or Discharge to Affect Grid Reliability?

Excessive load growth or generation penetration (e.g., renewable resources) can instigate grid thermal and voltage violations. For grid violations that are predominantly driven by excess load, operating energy storage in a discharge mode is helpful, while for grid violations driven predominantly by excess generation, a charging mode is helpful. In a general setting with multiple grid thermal and voltage violations, one or multiple coordinated storage systems, some operating in discharge mode and others in charge mode, might be required to effectively address these grid reliability violations.

1.5 How Much Energy Capacity Is Required?

The obvious initial driver for the required level of energy storage capacity is the hourly profile of the line overloads. The higher the overloads, the higher the power that should be charged or discharged; while the longer the duration of the overloads, the deeper the energy reservoir need to be sized.

However, there is also a not-so-obvious driver. Energy storage, by its nature, is an asset with limited energy and, therefore, should have the opportunity to recharge if it has been called upon to discharge to address a grid need (or to discharge if it already charged to address a grid need). Recharging a storage system increases the loading of the grid and might trigger a grid violation if not performed when the grid can accommodate it. These reliability restrictions on the timing to recharge the storage system, in some cases, drive the energy capacity requirements higher. For example, if, during a consecutive five-day period, the hourly load forecast is high, requiring the storage system to discharge six hours each day to shave the load, but during the remaining hours of each day, the storage can only be recharged by half the energy amount that was discharged, then the storage capacity will have to be upsized to 3.0x the daily energy discharge needs. On the first peak day, the storage will enter the peak load period with 3.0x capacity and will enter the second day with 2.5x, the third day with 2.0x, the fourth with 1.5x, and the fifth with 1.0x. Energy storage should be designed to have enough energy capacity to either ride through the overload duration or to provide ample time until grid operators switch loads or redispatch generation.

1.6 Can Storage System Increase the Grid Transfer or Hosting Capacity?

Unlike conventional T&D lines, energy storage does not create new grid capacity. Instead, it exploits the hourly profile of load and renewables to maximize the utilization of existing grid capacity. Transmission systems are typically half-loaded in anticipation of outages to avoid the risk of overloading transmission circuits before the generation can be redispatched. This is due to the inherent limited availability of automatic controls in the transmission grid such as line switching. Energy storage provides the requisite controls to increase the grid loading or transfer limits up to but not exceeding the thermal limits. Similarly, in distribution networks, energy storage can shave peak load, mitigate the effects of reverse flows and intermittency, and thus increase the hosting capacity.

1.7 When Can a Storage System Provide Other Services?

Grid violations can occur during normal or intact (N-0) grid operation, but more prevalently they occur under grid contingencies (e.g., N-1 or N-1–1). These grid violations are frequently seasonal (e.g., summer-peak or spring off-peak), and thus, the need for storage assets to provide grid services is typically infrequent and seasonal. For example, a winter-peaking utility might project a few hours or days during the winter season when the utility's reliability standards will be violated. The MW and MWh storage-capacity requirements to mitigate grid violations will vary hourly, depending on the degree of these violations. The operation of the storage asset can be optimized to provide market services such as capacity or ancillary services and/or to provide services to the local host such as backup power whenever the full storage capacity is not required for grid reliability. This value-stacking capability is an important feature of storage systems that can help to offset its cost. However, in terms of priority, reliability is the primary function and takes precedence, while revenue stacking is a secondary function that can be exploited when the reliability needs permit.

1.8 Analysis Techniques When Planning Energy Storage

The increasing variability and unpredictability of power flows in the T&D grids, instigated by the proliferation of distributed energy resources (DERs) and variable renewable resources (VERs), increases the complexity of grid planning and necessitates the development of more advanced analysis tools and processes. The traditional capacity planning that assumed dispatchable centralized resources and unidirectional flows on distribution grids is giving way to performance planning that seeks to assure successful integration of DERs and VERs while adhering to reliability standards. The traditional practice of examining grid performance at a limited number of operating snapshots (e.g., summer-peak or spring off-peak) is no longer adequate, and in many cases, a more complex hourly (e.g., 8760) series power flow and contingency analysis is required. The variability of renewable profiles and load profiles and their uncertain alignment may require probabilistic analysis such as Monte Carlo simulations wrapped around the 8760 time-series analyses. Furthermore, the dynamic interactions between inverter-based resources (IBR) on the transmission and distribution grids following a system event on either side are increasingly driving the need for integrated T&D analysis.

On one hand, unlike traditional T&D lines, which work passively without time limits and without active controls to increase transfer capacity, energy storage provides a similar core functionality, but for a limited time window, and requires active monitoring and control to coordinate and manage its real and reactive outputs and state of charge. On the other hand, the footprint, modularity, and potential movability of energy storage is a significant advantage, especially when deployed in the vicinity of load centers and for edge-of-grid applications. The proper analysis of energy storage 'as an alternative or a complement to traditional wire solutions should explore not only the core functionality but also the flexibilities and vulnerabilities inherent in each technology and supporting systems and their ability to de-risk capital investments in the face of increasing levels of uncertainty.

Six key aspects are critical to the successful planning of energy storage as grid assets:

1. Siting
2. Sizing
3. Revenue stacking
4. Techno-economic lifetime comparative analysis
5. Control system signals and relay tuning
6. Protection system design and settings

A methodology is proposed in this Element for each of the first four aspects along with applications and case studies. Siting and sizing analyses methods and formulas are presented in Sections 2.1 and 2.2 respectively for transmission grid applications. This is followed in Sections 3.1–3.4 by representative examples of the use of siting and sizing methods in market efficiency, grid reliability, renewable integration, and sub-transmission interruptions and load backup applications. Section 4 presents methods and an example of siting and sizing storage for distribution grid applications. A techno-economic comparative analysis methodology is presented in Section 5, while Section 6 presents techniques of valuing the optionality of energy storage. Modeling of energy storage in power system analysis tools is provided in Section 7 for steady-state, stability, and transient analysis applications. Additionally, three case studies are provided in Section 8 that illustrate the applications of energy storage as NWA; The first case study is focused on a transmission reliability application; the second on a distribution reliability application; and the third on the integration of wind assets into a transmission grid. A detailed presentation of revenue stacking is presented through the second case study in Section 8.2.

1.9 Transmission Reliability Criteria

Transmission grid planning is mandated to comply with minimum performance standards, such as NERC TPL-001–4 in North America. The performance requirements ensure that within the planning horizon, the Bulk Electric System (BES) will operate reliably over a broad spectrum of system conditions and following a wide range of probable contingencies as summarized in Table 1.1. Contingency classification per the NERC TPL-001–4 standard can be summarized as follows: P0 is intact system (N-0); P1 is a single-element failure (circuit, generator, transformer, shunt device); P2 is also a single-element failure

Table 1.1 NERC TPL-001–4: P1–P7 categories

	Type	**Initial Loss**	**Contingency**
P0	N-0	-	-
P1	N-1	-	Gen, T Ckt, Trafo, Shunt
P2	N-1	-	Line Section, Bus, Breaker
P3	N-1–1	Generator	Gen, T Ckt, Trafo, Shunt
P4	N-x	-	Stuck breaker (Gen, T Ckt, Trafo, Shunt, Bus Section)
P5	N-x	-	Delayed fault clearing due to relay failure (Gen, T Ckt, Trafo, Shunt, Bus Section)
P6	N-1–1	T Ckt, Trafo, Shunt	T Ckt, Trafo, Shunt
P7	N-x	-	Common Structure (2 Ckts)

NOTES:
N-0 = Normal system state
N-1 = Single contingency
N-x = Multiple contingencies
Ckt = Circuit

(line section, bus, breaker); P3 is a loss of a second element after a period of losing a generator (N-1–1), P4 is a multiple-element loss (stuck breaker), P5 is also a multiple-element loss (delayed fault clearing due to relay failure); P6 is a loss of a single element (line, transformer, shunt) followed by a loss of another single element (N-1–1), and P7 is a loss of multiple elements (common structure).

2 Siting and Sizing Methods for Transmission Applications

Transmission grids are operated well below their thermal capacities to ensure the security of supply deliverability even under a wide range of potential equipment outages and supply interruptions. Unlocking the latent underutilized capacity of the grid while ensuring system security has a great economic value [3].

Transmission systems are designed to interconnect generation and load areas and transfer energy efficiently while meeting a stringent set of reliability criteria [4]. Over decades, transmission networks have expanded mainly by utilizing passive conventional systems known as "wires solutions" (i.e., transmission lines, transformers, switchgear), which require an operational philosophy of keeping networks almost half-loaded in a preventive posture in anticipation of potential contingencies that push systems toward their thermal, voltage, or stability limits. Networks are restricted 100% of the time from reaching their full capability in transferring power from low-cost production regions to high-cost load centers due to the risk of events that typically last less than 2% of the time. These prudent operational restrictions, dictated by the capability of the conventional systems that make up transmission networks, lead to congestion costs that can have significant economic impacts on many stakeholders [3] [5] or hinder public policy goals of integrating renewable clean energy resources.

Energy storage offers many potential benefits to transmission and distribution (T&D) systems due to the ability of modern power electronics, and some electro-chemistries, to rapidly change from full discharge to full charge modes, or vice versa [2]. These characteristics have led to increasing interest in utilizing energy storage to economically unlock the inherent transmission or distribution grid capacity. Strategic siting and sizing of storage resources will allow the operators to load the grid above its contingent capacity and closer to its intact or normal system capability (Figure 2.1), and to utilize the storage resources to absorb or inject power post contingencies up until a system re-dispatch is invoked.

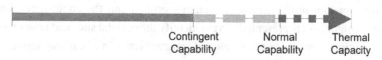

| Contingent | Normal | Thermal |
| Capability | Capability | Capacity |

Figure 2.1 Indicative operating limits in a transmission grid

This section provides a theoretical basis for optimizing the sites and sizes of energy storage that can increase the reliable transfer capacity across a transmission interface, flow gate, or boundary. These methods and formulas are original work of the author and have been utilized extensively in the conduct of many NWA projects. Several publications on siting and sizing energy storage have focused on optimizing the economics of renewable variable energy resources [6]–[9] or system frequency response [10], while others [11]–[16] have focused on generalized optimization methods for alleviating grid constraints or spatial-temporal energy arbitrage. In contrast, the approach presented in this section provides many advantages:

- Highly scalable to large power systems.
- Focuses on transmission or networked distribution grid security under all planning contingencies (P1–P7) [4].
- Takes as input the familiar representation of the system in terms of a power flow model and the set of contingencies and monitored facilities without the need to precalculate transfer capacity limits.
- Can be performed by transmission planners by manipulating existing power flow and security analysis tools.
- Provides valuable locational insights that enable an iterative planning process using storage-only or hybrid solutions encompassing combinations of conventional wire solutions and storage solutions.

2.1 Siting Analysis

The power flow on a line or through a transformer is influenced by the power injections at all buses through Power Transfer Distribution Factors (PTDF), and also by outages of other lines through Outage Transfer Distribution Factors (OTDF). Discharging an energy storage at a particular bus will influence the flows on all lines either in a decreasing manner or an increasing manner depending on its location and the prevailing direction of power flows.

The optimal site for an energy storage is one that reduces the flows on all congested lines under all contingency scenarios. In practice, several sites might be able, or required to mitigate a group of grid congestions, and thus a ranking of candidate sites is necessary in terms of their effectiveness to relieve congestion. The analysis is influenced by the grid structure (topology and equipment ratings), loading profile, renewable energy profile, and the contingency list. A robust algorithm is presented next to identify the optimal sites and rank them. The resulting siting index can be plotted geographically in a heat map format to guide the site selection process.

The transmission line overloads in a transmission grid with a set of injection (generation and load) nodes denoted as i, a set of congested lines denoted as k, and a contingency list of equipment outages denoted as j can be expressed in a linearized (i.e., DC power flow) format as:

$$Overloads = MAX_j \left\{ \sum_k \mu_k \left(\sum_i h_{ki}.P_i + L_{kj}.F_j \right) \right\}, \qquad (2.1)$$

where pre-outage power flow on line j (F_j) as a function of power injection at bus i (P_i) is

$$F_j = \sum_i h_{ji} P_i; \qquad (2.2)$$

- μ_k is a scaling factor representing the relative importance of line k, or the marginal cost of congestion of line k.
- $h_{ki} = \frac{\partial F_k}{\partial P_i}$ is the power transfer distribution factor, or, specifically, the change in power flow on line k due to injection at bus i.
- $L_{kj} = \frac{\partial F_k}{\partial F_j}$ is the outage distribution factor, or, specifically, the change in power flow on line k due to outage of line j.

The sensitivity index of the system overloads to a power injection at bus i can be derived as

$$\frac{\partial Overload}{\partial P_i} = MAX_j \left\{ \sum_k \mu_k \left(h_{ki} + L_{kj} \cdot h_{ji} \right) \right\}. \qquad (2.3)$$

A variation on this formula is to consider the probability of each contingency in lieu of the worst-case approach embodied by the MAX function. Another variation is to set μ_k as a function of the percentage overload of the monitored line.

For technologies connected in a shunt configuration to a grid bus, such as generators, loads, and energy storage, the locations with the largest negative (or positive) sensitivity index are the most influential at reducing the congestion when the technology is operated in a discharge (or charge) mode. The analysis can either be repeated for each chronological hour or performed on a select set of snapshots of loading profiles. The optimal sites are ranked by sorting the annualized overload sensitivities for each of the potential sites as calculated using the formula, Eq. (2.3), in ascending order, with the highest being the best site for locating an energy storage. This formulation does not require any optimization algorithm to calculate. The scaling factors μ_k are relative and can be taken as uniform for all lines, or skewed to reflect the level

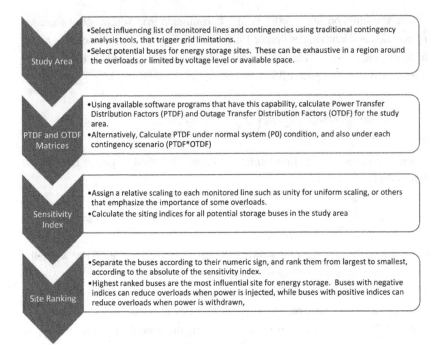

Figure 2.2 Siting analysis flow chart

of line loading or, if available, the marginal congestion cost of each line. Reliability studies will use the uniform scaling, while market efficiency studies can be skewed.

Figure 2-2 outlines a flow chart for the siting analysis.

2.2 Sizing Analysis

The optimal size of a group of energy storage systems can be calculated using a linear program (LP) that seeks to minimize the maximum overload under all contingency scenarios while simultaneously reducing the combined costs of these systems.

The objective of adding energy storage to relieve grid congestion is to allow the dispatcher to load the transmission grid, during normal operation, up to its normal (i.e., N-0) limits, and for the storage systems to remain ready to offset any potential overloads that may result after a contingency event.

If the flow on a path k is denoted by F_k^{post} during a post-contingency event and F_k^{pre} prior to the contingency event, and if the injection at bus i is denoted by P_i^{post} during post-contingency and P_i^{pre} prior to the contingency event, then

$$F_k^{post} = \sum_i h_{ki}.P_i^{post} + L_{kj}.\sum_i h_{ji}.P_i^{post}, \tag{2.4}$$

and the change in line k flow resulting from a line j outage is

$$\Delta F_k^j = F_k^{post} - F_k^{pre} = \sum_i h_{ki}.\left(P_i^{post} - P_i^{pre}\right) + L_{kj}.\sum_i h_{ji}.P_i^{post}. \tag{2.5}$$

The energy storage sizing problem can be formulated as a linear program (LP), the aim of which is to minimize the largest increase in overload resulting from all contingency events, as follows:

$$Minimize\ (\max_j \sum_{k\ \epsilon\ congested\ lines} (\ \mu_k.\Delta F_k^j) + \sum_{i\ \epsilon\ all\ buses} \propto_i \left(P_i^{post} - P_i^{pre}\right)),$$

$$\tag{2.6}$$

subject to the following:

- $P_i^{min} \le \left(P_i^{post} - P_i^{pre}\right) \le P_i^{max}$; for all buses i where energy storage is contemplated, where P_i^{max} and P_i^{min} are the discharge and charge power rating limits of a storage system at bus i (in part due to substation capacity limits).
- $\sum_i P_i^{post} = 0$, to preserve the power balance in the post-contingency state (optional).
- $\propto_i$ is the marginal cost of power generation or storage at node i.

This formulation is general in nature and allows the grid planner to optimize the size of single, or a set of coordinated energy storage systems. It also allows balanced or unbalanced storage solution strategies, where a balanced strategy will require that the storage systems charge and discharge in a manner that has a zero-net injection into the power system and, thus, avoid any interaction with the energy market even during a brief period following the occurrence of a rare contingency. The solution of the LP is very fast (e.g., seconds). The formulation, Eq. (2.6), can be modified to consider reliability constraints with a few additional constraints in the LP formulation.

The siting and sizing methodologies presented in this section are applicable when the grid violations are of a thermal line overload nature. However, they do not extend to situations where the primary grid concerns are voltage related.

3 Energy Storage as a Transmission Asset (SATA)

This section explores four key applications of energy storage, namely, reliability, market efficiency, renewable integration, and load backup. The first three applications are similar in their core desire to alleviate thermal or voltage violations under P1–P7 contingency events, while the fourth aims to optimize the sizing and operation of a microgrid supply. However, all four applications differ in the sizing optimization of energy storage and the types of technical and economic analyses that are required.

3.1 Market Efficiency Application

Transmission congestion is frequently encountered in post-contingencies, such as the loss of a transformer or a line, in what is labeled as N-1 congestion. However, it can also occur in normal system conditions if the daily load peaks cause a line to overload, in what is labeled as N-0 congestion.

To illustrate the concept of using energy storage to resolve N-1 congestion, Figure 3.1 shows a power system with three parallel lines, each rated 0.5 p.u. (per unit), delivering energy to serve a 1.5 p.u. load center. Two power plants can supply the load; one is remote from the load and economic, while the other is near the load but expensive. The grid can deliver the full 1.5 p.u. power requirements to the load from the remote generator if the system is operated with no regard to security, as shown by the graph in Figure 3.1. However, taking N-1 contingencies into account, system operators will dispatch the nearby power plant to mitigate the possibility of line overloads, should one of the three parallel lines open. So, ideally, the system operators would prefer to dispatch the system as N-0, but due to security concerns, they are forced to operate in N-1 security-constrained dispatch mode.

Energy storage can be introduced at the appropriate locations within the grid to operate only immediately after a contingency occurs, allowing the grid operator to dispatch the system in a manner similar to N-0 mode all the time, and only re-dispatch the generation resource post N-1 events. This operational philosophy has the potential for considerable cost savings in areas where traditional line solutions are not feasible or are expensive. Energy storage effectively removes from consideration contingencies that limit the flow on the transmission interface. Figure 3.2 shows the system operating in

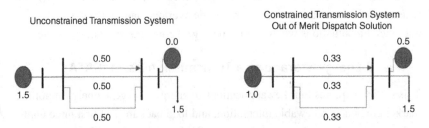

Figure 3.1 Out of merit dispatch to mitigate N-1 transmission congestion

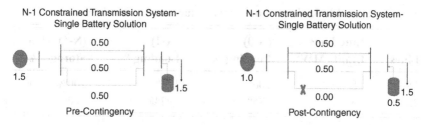

Figure 3.2 Energy storage for N-1 congestion relief

N-0 mode, and when one of the three lines opens, the energy storage will discharge to meet load requirements, giving system operators an adequate time period to reduce the remote generation level and start a local generation resource to bring the system into balance. With energy storage, the system can operate economically all the time, while meeting its reliability objectives during contingency events.

The following IEEE 30 bus example (shown in Figure 3.3) will illustrate the technical feasibility of using energy storage to mitigate the post-contingency overloads. Following the outage of the highlighted line segment 12–15, the flow on two lines (15–23 and 25–27) approaches alarming levels. The siting algorithm suggested several sites to place a storage system, while the sizing algorithm optimized the battery MW size. The battery MWh depends on the operational flexibility of re-dispatching the system and is typically 30–60 minutes. The post-contingency flows are restored to within the line rating limits as illustrated in Table 3.1.

3.1.1 Indicative Economic Analysis of Congestion Relief Using Energy Storage

For a congested line (flow gate, interface, or boundary), the following example will serve to illustrate the indicative economics of utilizing energy storage as non-wire solutions for N-1 contingency-constrained grids.

Assume the following conditions:

- The flowgate is constrained 15% of the time (i.e., 1,350 hours in a year).
- Average congestion cost during congestion hours is $10/MWh.
- Load affected by relieving flow gate congestion is six times the size of the line overload.
- Transmission line solution length is 23 miles.
- Transmission line capital cost per mile varies between a low of $2 million and a high of $5 million, with an average of $3.5 million.

Table 3.1 Impact of energy storage on line overloads

Lines	Line Flow Limit (MW)	(N-0) Loading %	(N-1) Loading %	(N-1) After Storage %
15–23	16	100%	123%	100%
25–27	16	85%	92%	80%

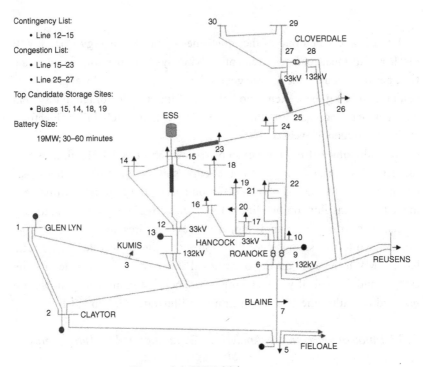

Figure 3.3 IEEE 30 bus system

- Achieved level of flow gate capacity increase with energy storage is 100 MW.
- Annual O&M is 2% of capital investment.
- Transmission line permitting and construction period varies between a low of one year and a high of five years, with an average of three years.
- Battery-based storage system installed cost ranges between a low of $450/kWh and a high of $650/kWh, with an average of $550/kWh.
- Storage annual O&M cost varies between a low of $10/kWh to a high of $18/kWh, with an average of $13/kWh.
- Storage capacity is 1 hour, and its lifetime is 15 years.
- Weighted average cost of capital is 8%.

Although these assumptions were selected for illustrative purposes, most of them are reasonable and expected to persist over the upcoming decade. Some assumptions are system specific such as congestion hours per year and flow gate capacity increase level, while others like storage cost are highly variable and expected to decline over time.

Figure 3.4 compares the annualized cost of grid congestion to the cost ranges of the two alternative solutions: transmission lines and storage system. This example illustrates that the relative economics depends on the prevailing situation in the region. Storage-based solutions can be economical when compared to transmission lines if the line cost is high, its permitting period is long, or if the storage cost is low or when the storage can derive secondary revenues. As storage systems continue their cost reduction road map, their economic advantage is expected to increase.

3.2 Grid Reliability Application

The flows and voltages of transmission grids are highly influenced by the load and renewable generation levels. In places where the load increases to a level that causes a violation of prevailing thermal or voltage standards under normal or N-1 conditions, an optimally placed storage system might provide the best economic solution instead of building lines or adding reactive compensation solutions. Key issues to consider in the design of the non-wire solution are the nature of the limiting conditions in terms of voltage and thermal ratings, and the ability to relieve the voltage part of the grid violations through adjustments of cap banks or on-load tap changers (OLTCs) to reduce the need to upsize the storage inverter rating, in addition to applicable planning criteria of setting a reasonable safety margin on the thermal flow limits. Other key design criteria are the site selection, the storage reactive power capability, and the ability of the storage system to recharge before it is needed.

A simplified two-area system as shown in Figure 3.5 is utilized to quantify the benefits and costs of relieving the reliability limits of an interconnection using an energy storage investment in lieu of a conventional transmission line expansion.

The existing transmission lines are fully subscribed during peak load periods, requiring a grid expansion. The grid expansion can be performed using a conventional transmission line upgrade or through a non-wire alternative including, among several options, integrating a storage system downstream of the congested lines, within the transmission grid of Area B. System expansion using non-wire alternatives to relieve transmission congestion, for reliability reasons, is most justified when the daily consecutive congestion hours are limited, and the transmission upgrades are expensive.

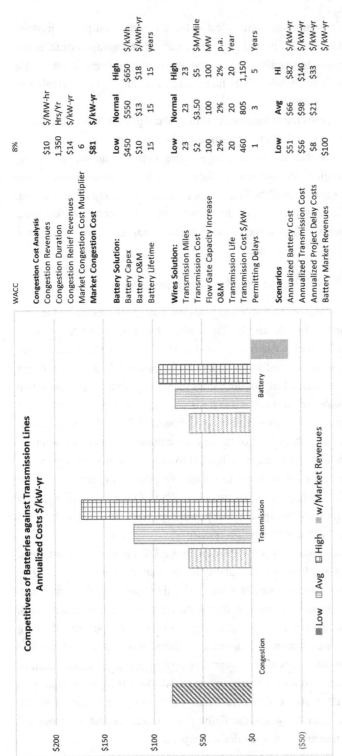

Figure 3.4 Comparative cost of storage solution to a conventional line solution

WACC	8%			
Congestion Cost Analysis				
Congestion Revenues	$10			$/MW-hr
Congestion Duration	1,350			Hrs/Yr
Congestion Relief Revenues	$14			$/kW-yr
Market Congestion Cost	**$81**			**$/kW-yr**
Battery Solution:	**Low**	**Normal**	**High**	
Battery Capex	$450	$550	$650	$/kWh
Battery O&M	$10	$13	$18	$/kWh-yr
Battery Lifetime	15	15	15	years
Wires Solution:	**Low**	**Normal**	**High**	
Transmission Miles	23	23	23	
Transmission Cost	$2	$3.50	$5	$M/Mile
Flow Gate Capacity Increase	100	100	100	MW
O&M	2%	2%	2%	p.a.
Transmission Life	20	20	20	Year
Transmission Cost $/kW	460	805	1,150	
Permitting Delays	1	3	5	Years
Scenarios	**Low**	**Avg**	**Hi**	
Annualized Battery Cost	$51	$66	$82	$/kW-yr
Annualized Transmission Cost	$56	$98	$140	$/kW-yr
Annualized Project Delay Costs	$8	$21	$33	$/kW-yr
Battery Market Revenues	$100			$/kW-yr

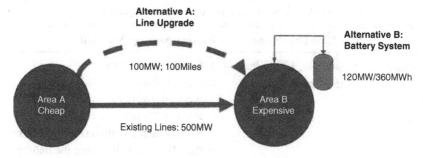

Figure 3.5 Non-wire alternative in a two-area system

The analysis shown in Table 3.2 quantifies the comparative economics of a transmission line upgrade and a non-wire storage-based upgrade alternative. We will consider a congestion profile with two daily periods during the three peak months of the year, exceeding the tie line capacity by 70 MW from 12–2 PM, and 100 MW from 7 to 10 PM. The distance between Areas A and B is taken as 100 miles. The financial analysis considers, for both technology alternatives, the optimum sizing, the capital and O&M costs, the asset life, and energy losses.

The comparative economics (Table 3.3) between a transmission line and a non-wire storage solution over 20 years, assuming a battery cost of $500/kWh and a transmission line cost of $1 million USD per mile, with and without providing frequency regulation services, shows investment returns that are marginal without access to frequency regulation services. However, as the battery storage cost decreases and/or the cost of building transmission lines increases, the relative economics of energy storage improves substantially, as shown in Figure 3.6. For a required investment internal rate of return (IRR) of 8%, the battery cost cannot exceed $420/kWh without any secondary revenues and $630/kWh with frequency regulation secondary revenues.

3.3 Renewable Integration Application

The integration of renewable resources such as solar and wind (onshore and offshore) can instigate grid reliability or market efficiency concerns during a few or many hours of the year. Additionally, the intermittency of these resources at high levels of penetration will require added flexibility in the thermal fleet to mitigate the impacts of power ramps and low inertia on system frequency excursions. The steady-state issues of grid reliability and market efficiency can be studied in the same manner as previously discussed in Sections 3.1 and 3.2.

Table 3.2 Study parameters and assumptions

	New Line Expansion	Battery Storage Solution
Sizing	132-kV line, 100 miles, carrying 100 MW	120 MW / 360 MWh Designed to offset the full congestion while allowing for battery degradation at 2% per year for 10 years. Battery augmentation (60 MWh) in the 11th year will restore the battery to original ratings and extend its life to 20 years.
Capital Cost	$1 million USD per mile for a total of $100 million	$500/kWh for a total of $180 million
O&M Cost	1% of capital cost escalating at 1.5% per year	2.5% of capital cost escalating at 1.5% per year
Losses	5%	10% (AC-AC roundtrip efficiency of 90%)
Asset Life	20–40 years	15–20 years

Table 3.3 Investment economic in storage as NWA

Incremental Investment Economics of Energy Storage	Payback Years	IRR %
without Frequency Regulation revenues	13.4	3.9%
with Frequency Regulation revenues	4.5	21.5%

Storage systems that are deployed in the grid for transmission expansion can also be simultaneously utilized to enable the system to incorporate larger penetrations of intermittent renewable resources. Dependent upon the hourly and seasonal profiles of wind and solar production and their alignment with load consumption profile, storage systems may have latent capacity to provide grid services in addition to congestion mitigation such as:

• Inertial and primary frequency response
• Ramp rate control
• Short circuit strength

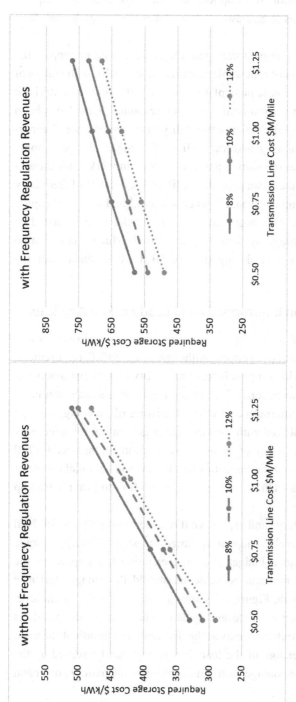

Figure 3.6 Contours of constant battery investment IRR as a function of battery and transmission line costs

This feature becomes important in projects that have fast regulation requirements, like island systems (e.g., Hawaii), or in places that integrate electrified
transportation.

The power rating of energy storage that may be required to support the
integration of high levels of renewable resources is typically in the range of
10% to 40% of the renewable nameplate capacity, sufficient to control the
ramp rates and provide the required frequency response. This depends on
several factors including the existing flexibility in the conventional generation fleet, the predictability and variability of weather, the geographic
footprint, and the market scheduling periodicity. Recent work demonstrates
that battery storage systems can provide six times the amount of frequency
response benefits as comparably sized gas generators,[1] and thus can be quite
effective economically. This high multiple reflects the relative speed of
response of battery storage systems to that of diesel-fueled combustion
engines in delivering power during the crucial 1–10 seconds after the
onset of an event.

3.4 Sub-Transmission Interruptions and Load Backup Application

Sub-transmission radial lines expose customers to prolonged outages that
may require over six hours to restore. Although not a NERC requirement,
utilities desiring to provide a higher level of reliability to their customers can
either build another conventional line connecting to an alternate source or
can design a non-wire alternative solution consisting of an energy storage
facility potentially combined with a solar or other generation resource. Key
issues to consider in the design of the non-wire solution are the need for full
or partial load backup, and the impact of the diurnal and seasonal nature of
renewable sources on the ability of the storage solution to provide backup
services.

The solution will be large and expensive if required to supply the full load
requirements. Alternatively, one can optimize the size of energy storage
(MW and MWh) within a budget and then examine the impact on the
resulting backup level that can be achieved should the outage start at a
particular hour and month. Figure 3.7 shows a sample of a minimum level
of load backup achieved by a storage system. The hourly load profile is
analyzed assuming the outage occurs at the first hour and month of the year,
and the resulting percentage of the load energy that can be served by the
storage system during the outage with consideration to the battery power and

---

[1] USTDA-funded project: Feasibility Study for Large Scale Energy Storage Systems in Brazil,
Colombia, and Mexico: Task 10 Report (Public).

Minimum Backup Level

min Load Backup when Outage Event Occurs at (month, hour)

Min Col ▼								Outage Month			
Ro ▼ 1	2	3	4	5	6	7	8	9	10	11	12
0 37%	49%	51%	85%	92%	85%	69%	68%	75%	61%	59%	35%
1 36%	47%	55%	84%	94%	92%	69%	67%	74%	66%	64%	35%
2 36%	50%	56%	84%	92%	99%	68%	65%	72%	72%	68%	36%
3 37%	50%	59%	85%	92%	100%	66%	63%	71%	79%	73%	36%
4 38%	52%	60%	90%	93%	100%	65%	60%	70%	80%	77%	37%
5 39%	55%	61%	91%	94%	100%	63%	58%	67%	83%	78%	39%
6 41%	58%	63%	93%	91%	100%	61%	55%	66%	83%	78%	40%
7 43%	63%	65%	95%	87%	100%	58%	52%	62%	83%	79%	43%
8 46%	73%	65%	96%	86%	100%	55%	50%	59%	83%	77%	45%
9 47%	79%	66%	96%	82%	96%	52%	48%	55%	81%	75%	46%
10 48%	78%	68%	94%	80%	90%	49%	46%	51%	78%	73%	46%
11 47%	75%	65%	88%	76%	83%	47%	46%	49%	75%	71%	46%
12 47%	73%	62%	85%	74%	77%	46%	45%	48%	72%	67%	45%
13 45%	65%	59%	80%	66%	71%	45%	45%	48%	69%	64%	43%
14 44%	58%	55%	79%	59%	66%	45%	45%	49%	63%	63%	41%
15 43%	57%	53%	79%	58%	63%	46%	46%	50%	59%	63%	40%
16 42%	54%	51%	75%	60%	60%	46%	48%	52%	57%	59%	39%
17 39%	51%	48%	74%	64%	58%	47%	48%	51%	56%	56%	38%
18 39%	53%	48%	77%	68%	59%	48%	49%	53%	58%	56%	38%
19 39%	53%	48%	80%	71%	60%	50%	51%	57%	58%	57%	38%
20 39%	54%	49%	82%	75%	64%	53%	55%	61%	59%	57%	37%
21 39%	54%	49%	84%	79%	68%	56%	58%	65%	59%	58%	37%
22 38%	54%	49%	85%	82%	72%	59%	61%	67%	60%	58%	37%
23 37%	52%	48%	85%	86%	76%	63%	65%	70%	59%	57%	36%

(Left vertical axis label: **Outage Start Hours**)

Figure 3.7 Load backup heatmap by hour and month

energy limitations is tabulated for each day of the first month. The minimum backup (or alternatively, average or 90th percentile) is calculated and tabulated in the upper-left cell of the table. The results for an outage occurring in the second hour and first month are tabulated in the first left-side column in the second row, and so on. We can see that if the outage were to occur at hour beginning 10 in June, the energy storage system will be able to supply 90% of the load requirements, as highlighted in Figure 3.7.

The storage system can be optimized to meet budget constraints based on the temporal probability and consequence of the outages. For example, an optimization process can be wrapped around this type of analysis to select storage size (MW and MWh) that minimizes an objective function subject to meeting specific constraints, including a weighted measure of reliability based on probability and consequence of outages, and storage capital cost. For example, Figure 3.7 indicates that if outages occur predominantly during the spring months, the storage size requirement to provide full coverage will be quite smaller than if outages occur during winter months.

4 Siting and Sizing Methods for Distribution Applications

The distribution system has traditionally transported energy from the transmission system to individual residential, commercial, and industrial consumers and met reliability and power quality standards of service. Increasingly, the distribution system is also integrating distributed generation (DG) resources that require careful management of the ensuing voltage fluctuations and coordination of the protection system to tolerate the effects of bidirectional power flows. The typical distribution system topology is anchored by distribution substations, each serving several medium-voltage feeders that traverse the utility's service territory. Most distribution feeders in the US are radial. Many are equipped with reclosers to enable sectionalizing the feeder after an outage and providing alternative sources to improve delivery reliability. In dense urban areas, the distribution feeders can be meshed and are operated as an integrated network to provide a higher level of delivery reliability.

Two key applications of energy storage in the distribution grid are feeder deferral and reliability improvement. Important issues are the siting and sizing of energy storage that can effectively defer feeder upgrades or improve service reliability.

4.1 Feeder Peak Shaving Application

To illustrate the competing considerations in the siting of energy storage, a representative radial feeder serving multiple loads along its path is shown in Figure 4-1. The feeder starts at a distribution substation and serves ten load pockets along its way (L1–L10). The current flowing on the feeder changes after each load center. The substation serves all the load, and thus the current flowing through its service transformers is the highest. The current drops after each load center until it reaches zero at the far end of the feeder. The upper staircase line depicts the current flowing in each segment of the feeder. Assuming all feeder segments have the same ampacity, and assuming the feeder-current carrying limit is shown by the horizontal dashed line, then it is clear the feeder is overloaded starting from the substation to the load labeled L3, or between the points labeled 1 and 2 on the feeder current graph.

Discharging energy storage can alleviate the feeder overloads, but only if located appropriately. Placing the storage at the substation does not help since the same current will have to flow along the feeder to serve the load. Placing the storage system downstream of load L3 has the opportunity to reduce the feeder overload upstream of its location. If the storage size is sufficient, the overload along the entire feeder can be eliminated. In this example, the storage discharge should not be less than the maximum overload, which in this case occurs at the substation. How far downstream from load L3 can the storage be placed? The feeder current

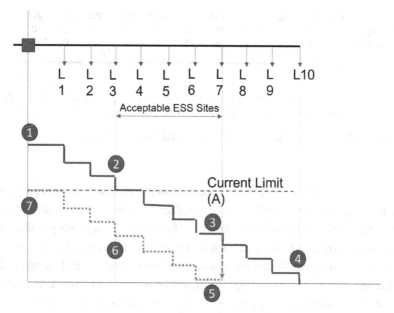

Figure 4.1 Representative distribution feeder with ten load branches

decreases further away from the substation, and therefore placing the energy storage further down the feeder risks creating back feed when the storage system is discharged. In this example, any place between loads L3 and load L7 is an acceptable location that does not cause a back feed and allows complete mitigation of feeder overloads. Placing the storage system just after load L7 will reduce the feeder current from the point labeled 3 to the one labeled 5 when the storage is discharged. Going back, the feeder current at the substation will drop from the point labeled 1 to the one labeled 7, just below the feeder current limit. If back feed does not cause any issues with the protection system, then the storage can be located further downstream from load L3 until its discharge causes an overload on the immediate segment upstream from its location.

The same methodology can be used when the feeder ampacity changes along its path, but care should be exercised to plot percentage overload along the feeder sections and then find single or multiple energy storage sites that can alleviate overloads.

4.2 Feeder Reliability Application

The previous example is expanded here to illustrate the use of energy storage to improve feeder reliability. Each feeder section may pass through a different terrain or have a different type of construction from other feeder sections,

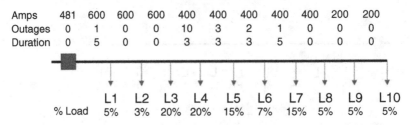

Amps	481	600	600	600	400	400	400	400	400	200	200
Outages	0	1	0	0	10	3	2	1	0	0	0
Duration	0	5	0	0	3	3	3	5	0	0	0

| | L1 | L2 | L3 | L4 | L5 | L6 | L7 | L8 | L9 | L10 |
| % Load | 5% | 3% | 20% | 20% | 15% | 7% | 15% | 5% | 5% | 5% |

Figure 4.2 Feeder section information – ampacity, load, outages, and duration

causing its reliability to differ. One section might encounter a wooded area, another might traverse a hilly area, and another might be a buried cable instead of a bare overhead conductor. To quantify the feeder reliability and then improve it using energy storage, each feeder segment is assigned two reliability indices, one to capture the expected number of outages per year and the other the average duration of outages. Each feeder section is assigned an ampacity rating.

Reclosers are a technology that is used to provide an alternative source from a nearby feeder. Energy storage is also another technology that can be sited and sized to improve feeder reliability. This example will explore a methodology to site and size energy storage to improve feeder reliability.

The expected energy-not-served (EENS) is one metric of feeder reliability that can be optimized by siting and sizing energy storage. A Linear Program (LP) can be formulated to site and size energy storage along the feeder and optimize the recloser(s) infeed current, to minimize the energy not served.

In the following example, the ampacity, number of outages per year, and the expected outage duration are shown for each segment of the 12-kV feeder. Additionally, the percent of the total feeder loading drawn at each load point is shown in Figure 4-2. The total load served by the feeder is assumed to be 11.3 MVA (or 546 amps). The overload at the substation headend is 13.5%. A recloser is connected just after the load L6 with a maximum capacity of 100 A that can provide an alternate source of power from a neighboring feeder.

Table 4-1 summarizes the results of the optimization and shows the required location and sizes of energy storage and recloser capacity for various levels of reliability requirements. In this example, no back-feed is allowed on any feeder segment. As the desired level of reliability is increased (by decreasing the acceptable level of ENS), more storage systems are located closer to each load center.

5 Techno-Economic Comparative Analysis Methodology

An NWA solution that includes an energy storage asset should be evaluated on the same basis as a conventional solution. However, there are differences

Table 4.1 Placement and size of energy storage to achieve an acceptable level of reliability characterized by EENS

Accepted Level of EENS (MWh-Yr)	Required Storage MVA	Required Recloser AMPs	Storage Placement and Size (MW)									
			L1	L2	L3	L4	L5	L6	L7	L8	L9	L10
291	0	100										
250	0.76	100									0.19	0.57
200	1.69	100								0.55	0.57	0.57
150	2.72	100					0.60	0	0.42	0.57	0.57	0.57
100	3.87	100				0.05	1.70	0	0.42	0.57	0.57	0.57
50	5.30	100				1.48	1.70	0	0.42	0.57	0.57	0.57
25	6.01	100				2.19	1.70	0	0.42	0.57	0.57	0.57
20	6.55	78			0.18	2.27	1.70	0	0.88	0.57	0.57	0.57
15	7.55	38			1.18	2.27	1.70	0	1.70	0.57	0.57	0.57
10	8.55	38			2.18	2.27	1.70	0	1.70	0.57	0.57	0.57
5	9.55	38			2.18	2.27	1.70	0	1.70	0.57	0.57	0.57
0	10.55	38	0.57	0.34	2.27	2.27	1.70	0	1.70	0.57	0.57	0.57

between the attributes of the two categories of solutions, including the permitting duration, the expected life, the asset management strategy and costs, the operational risk, and the ability to provide additional services beyond the primary reliability function, which should be quantified and evaluated equitably.

The following proposed economic analysis methodology considers the lifetime costs of each solution including initial capital, operations and maintenance, capacity augmentation, asset replacement and renewal, and asset retirement. The analysis considers the revenue requirements of each solution as a regulated asset, and thus focuses on the cost to the utility customer. The evaluated cost can be adjusted up or down to account for additional value or revenue streams that a solution may provide.

The economic evaluation of the storage solutions as compared to the conventional T&D solutions requires:

- Lifetime modeling of the cost of each project from inception to retirement inclusive of project development activities and timeline, EPC (Engineering, Procurement, Construction), O&M (Operations and Maintenance), capacity fading management, replacement, disassembly, and recycling.
- Modeling of the relevant utility's capital structure including debt and equity ratios, costs, and tax rate.
- Proper regulated asset base (RAB) accounting, including treatment of asset depreciation.
- Useful asset life estimates: The conventional T&D solutions have an assumed book life of 40 to 60 years, while the energy storage technology has a typical useful calendar life of 10 to 15 years for lithium-ion technology and is further restricted by the usage profile and its impact on the life cycles.

The following methodology is suggested to compare the economics of various solution alternatives:

- The initial installed capacity of the energy storage part of a solution is upsized to mitigate the anticipated capacity fading throughout the asset life. For a nominal 2% annual degradation [17] of storage capacity, the installed storage MWh capacity can be upsized by 16% from the level required to address the system needs to have one augmentation at midlife.
- Straight-line depreciation of each asset is adopted for book accounting purposes. The tax depreciation methodology can vary depending on applicable rules. In the U.S., accelerated depreciation is used for tax depreciation.
- Due to the differences in asset life between the conventional asset (40–60 years) and the storage asset (10–15 years), the analysis is carried out over the

conventional asset life, and multiple battery investment/replacement cycles are modeled. Two approaches are considered for the comparative analysis; the first calculates the present value of each solution cost over the 40–60-year horizon, while the second focuses on the first life of the storage system (e.g., 15 years) and calculates the levelized real cost of the conventional solution over that period utilizing a real economic carrying cost, considering the utility's weighted average cost of capital and the inflation rate.

- The present value of the revenue requirement (PVRR) for each solution is calculated.
- The ratio of a storage solution's revenue requirements to that of the conventional solution is taken as the key metric of evaluation. When the ratio is below 1, the storage solution is deemed to be more cost effective to the ratepayers than the conventional solution.

Figure 5.1 shows a representative cost of an energy storage investment and the asset revenue requirements. The resolution in the graph is in quarters (i.e., 3-month periods). During the construction period, the installed cost will accrue over several months, and once the system is operational in the year 2027, the customers will start paying for its regulated cost, which consists of depreciation, return on net asset value, and operational costs. The O&M cost is depicted by the short columns. In the year 2034 the battery capacity is augmented to offset the fading, and in the year 2036, the inverters are replaced. After the battery end-of-life in the year 2041, the modules will be recycled, and new racks installed in addition to any required upgrades.

As shown in Figure 5.1, the most significant portion of capital cost occurs during the start of the project. Additional capital expenditures (Capex) relate to ESS augmentation schedules, inverter replacement, and battery replacement after the calendar life of the equipment. Increased ESS operational costs (Opex) post replacement pertains to the disposal of the battery.

Figure 5.2 compares the indicative revenue requirements of a storage system to those of a conventional line. The conventional line revenue requirements are declining smoothly as the asset is depreciated, while the energy storage revenue requirements vary during the various phases of the asset life. Lower customer cash flow (or regulated project revenue requirements) is observed when the battery participates in the wholesale markets to provide ancillary services and arbitrage whenever the reliability requirements permit.

The present value of the revenue requirements (PVRR) is a well-accepted metric to compare investments. If the ratio of the storage PVRR to that of the conventional solution is above one, then the storage is considered more expensive, and vice versa. The PVRR metric attributes higher value to investments

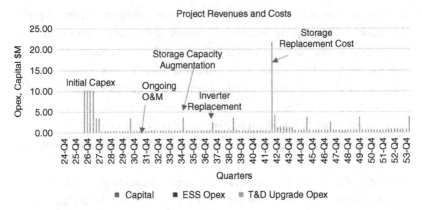

Figure 5.1 Representative cost elements of a storage system

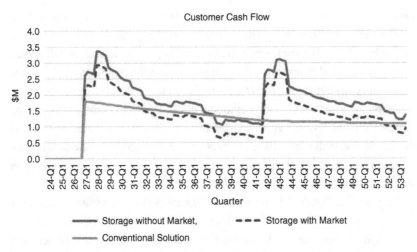

Figure 5.2 Comparative project revenue requirements of a storage system and a conventional line

that have cash outflow requirements (i.e., investments) in the future instead of the current time, and thus it is more beneficial to right-size the storage system at the beginning of its life and then plan on augmentations and replacements in the future as the application needs grow and/or the storage system capacity fades.

6 Energy Storage Optionality Valuation

Transmission and distribution planning is becoming more challenging in the face of increasing uncertainty in forecasting load growth levels. The needs assessment of conventional grid solutions of lines and substations depends on future (e.g., 10 years) projections of demand levels, which are highly influenced by customer adoption of distributed energy resources (DER). If a utility's 10-year forecast overestimates the grid need, new underutilized assets will be built that are costly to ratepayers. Alternatively, if the forecast underestimates the need, system security and customer experience will be jeopardized. Energy storage is increasingly recognized as an effective non-wire solution that is modular, quicker to permit, and expandable. This flexibility of an energy storage asset is a key attribute and advantage that can provide additional value to ratepayers beyond a comparably sized conventional solution, namely a risk mitigation from being stranded or an underutilized investment. Storage can be built based on shorter-term forecasts and can be upgraded as new demand forecasts demonstrate continued growth in the system need.

A real-options framework is introduced (Figure 6.1) to quantify an often-neglected value of a storage asset, namely flexibility. The concept is demonstrated using an indicative use case of a storage system that is planned to mitigate transmission reliability concerns. The formulation is also shown to address other planning uncertainties, including variability of stacked revenues due to price risks in the wholesale markets.

The traditional economic valuation methodologies used in storage studies, assume an all-or-nothing investment strategy, and do not account for the value of asset *optionality*, the concept that management can adapt their investment decisions over time as previously uncertain elements such as load development

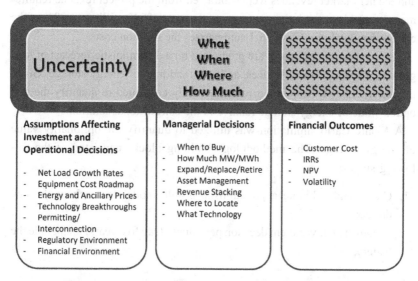

Figure 6.1 Modeling process of storage real option value

become known. A conventional transmission line tower, a right-of-way, or a substation can be sized with spare capacity to permit installation of an additional circuit or bay and thus provide planners with a right-sized solution to meet system needs in the short/medium term, and an option to expand the solution should the uncertain mid/long-term forecasts become more certain. Similarly, energy storage solutions are modular and can be designed to be expanded over time. Additionally, a storage solution's life is a fraction of a conventional solution asset, which lends itself to a multistep decision-making process.

6.1 Real-Option Example

A reliability study[2] for a load area within the U.S.'s Western Electricity Coordinating Council (WECC) showed the need for several transmission line upgrades ranging in voltage between 60 and 115 kV with a total revenue requirement (i.e., ratepayer cost) of $52 million. This solution is expected to address the load growth requirements over a 30-year period. An alternative hybrid solution consisting of a single energy storage combined with a line upgrade can provide a comparable performance over the same 30 years but has a higher lifetime revenue requirement of $65 M, or 125% of the conventional solution cost. However, the reliability requirements are clustered in the summer months, allowing the storage system to be utilized in the wholesale market outside the summer months. After quantifying the grid reliability requirements in terms of reserved MW and MWh capacity for each hour in a year, the remaining storage capacity was optimized in the wholesale market, and the net market revenues were subtracted from the project revenue requirements to yield a net solution revenue requirement of $43 million, or 82% of the conventional solution. Table 6-1 summarizes the solution costs.

These economic estimates are predicated on a deterministic forecast of load growth, wholesale market prices, and cost road map of energy storage. Given the uncertainty of these key drivers, the study continued to quantify the real option value of energy storage relative to the conventional solution.

A Monte Carlo simulation was utilized to quantify the optionality value of energy storage. The methodology building blocks consisted of the following stages:

1. Create and calibrate a probabilistic profile model for each of the key drivers.
2. Establish an investment decision periodicity (say five years) to upgrade the battery.

---

[2] Consulting study conducted by the author, but not published.

Table 6.1 Comparative financial analysis of storage and conventional solution

Comparative Customer Economics (Project Revenue Requirements): Customer Cash Flows over 30 Years ($M)		
	Discounted Cash Flow	**Nominal Cash Flow**
Conventional solution	52.0	140.2
Storage ESS without market participation	65.2	172.1
Storage as % of Conventional	*125%*	*123%*
Storage ESS with market participation	42.6	107.5
Storage as % of Conventional	*82%*	*82%*

3. Sample the probabilistic profiles of the key drivers at the investment periodicity.
4. Size (MW and MWh) the storage given the sample of the key drivers.
5. Quantify hourly reliability requirements of the storage.
6. Optimize the storage participation in the wholesale market.
7. Perform lifetime techno-economic evaluation of the cost of storage solution as compared to the conventional solution.
8. Reiterate steps 3 to 7 until the Monte Carlo convergence monitor stabilizes or a predetermined number of samples is reached.
9. Generate probability distribution plots of the solution need year, storage size, market participation revenues, and relative lifetime cost to the conventional solution.
10. Perform sensitivity studies around key assumptions.
11. Real option value is calculated as the difference between a base deterministic case and a sensitivity case that exercises an option. The options can be to size the initial storage system installation based on a five-year or a 15-year outlook, and to augment the storage in between as required. Another option is to exclude the battery from market participation during the reliability requirement season or to expand the participation to all months while respecting the hourly system reliability needs.

A few illustrative plots of the study data and results follow. Figure 6.2 shows a probabilistic load model that was used in the study, while Figure 6.3 shows the resulting installed battery size and the size of the battery augmentation, and Figure 6.4 shows the probability distribution of market revenues (in $/MW-yr) stemming from participation in the arbitrage and frequency regulation markets.

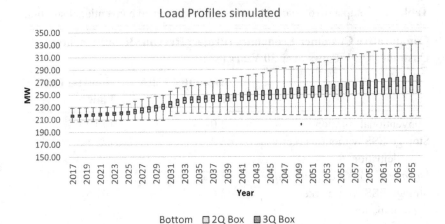

Figure 6.2 Probabilistic load model

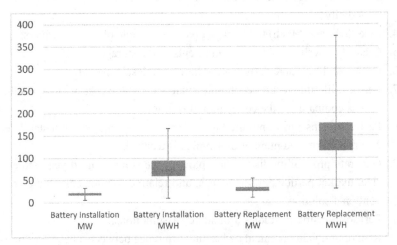

Figure 6.3 Range of battery initial size and augmentation size

The key output of the study is shown in Figure 6.5, which shows the probability distribution of the relative revenue requirements of the storage system to those of the conventional solution.

The relative cost of the storage solution ranges between 32 and 126% of a properly designed conventional T&D solution, with a mean relative cost of 73%, provided that the storage is optimally sized, based on load forecast samples. Under all sampled scenarios, the cost of the storage solution is 97% certain to be cheaper than the conventional T&D solution. This clearly shows that exploiting the flexibility of energy storage will, in 50% of the scenarios, save customers 27% of

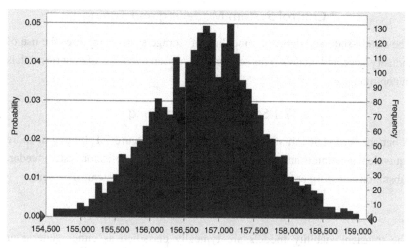

Figure 6.4 Probability distribution of market revenues in $/MW-year

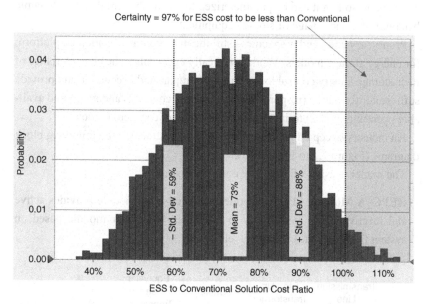

Figure 6.5 Probability distribution of relative revenue requirements of storage against conventional

the lifetime cost, with the potential to save them up to 68% if the future net-load increases are on the lower end of the forecast. However, it also has the potential to cost them 26% more if the net-load increases are on the higher end. The analysis shows that ratepayers would be better off with an optimized hybrid solution 97% of the time.

7 Control System Modelling and Tuning

The steady-state and dynamic modeling of storage systems requires the use of industry-accepted models. This section provides an overview of generic models in common use.

7.1 Steady State Modeling

A battery is modeled in steady state, as shown in Figure 7.1, by one or more equivalent generators and unit transformers, equivalent collector system feeder, substation transformer, and plant-level reactive support system, if present.

7.2 Transient Stability Models

The transient stability models are typically classified as either generic or proprietary:

Generic model – refers to a model that is standard, public, and not specific to any vendor, so that it can be parameterized to reasonably emulate the dynamic behavior of a wide range of storage equipment.

Proprietary and vendor specific – still the most commonly adopted platform for energy storage models. Vendors typically encapsulate the characteristics of their storage battery control loops within a wraparound executable and provide sufficient information (response curves, integral constants and gains, and available operating modes) to integrate them in dynamic system models.

An industry-accepted generic model [18] is illustrated in the following block diagrams (Figure 7.2 to Figure 7.5).

The model consists of three key components:

- **REPC_A Module:** Plant Controller Module – This module provides active and reactive power commands to the battery controller module based on system frequency and voltage levels.

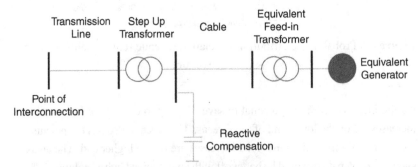

Figure 7.1 Equivalent model for steady-state analysis

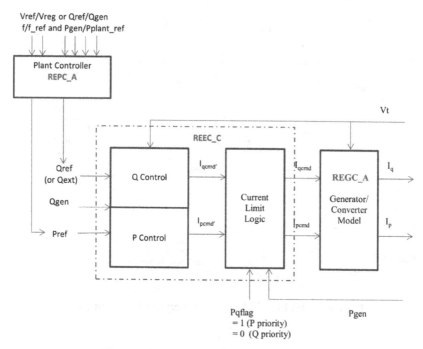

Figure 7.2 Generic storage dynamic model

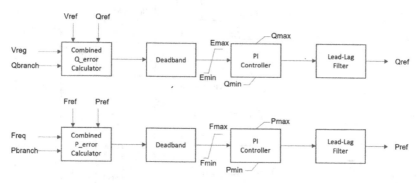

Figure 7.3 REPC_A Module – plant controller module

- **REEC_C Module:** Current Controller Module – This module translates the plant controller's instructions of real and reactive power to direct and quadrature current setpoints of the inverter while managing the battery state of charge.
- **REGC_A Module**: Converter Controller Module – This module represents the battery's converter (inverter) interface with the grid. It takes in the real and reactive current setpoints from the battery controller module.

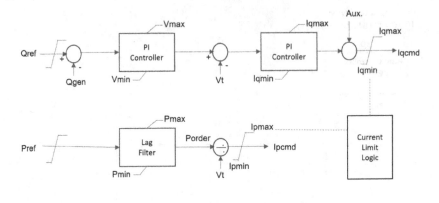

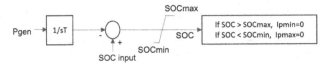

Figure 7.4 REEC_C Module – battery controller module

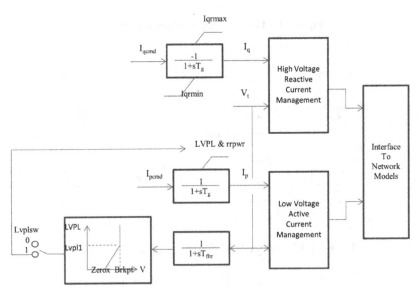

Figure 7.5 REGC_A Module – battery converter

7.3 Electromagnetic Transient EMT Model

A generic EMT model that is available in some common software programs [19], as shown in Figure 7.6, provides a generalized approach

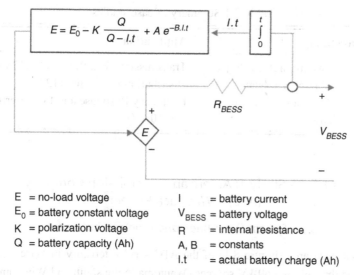

E = no-load voltage   I   = battery current
E_0 = battery constant voltage V_{BESS} = battery voltage
K = polarization voltage   R   = internal resistance
Q = battery capacity (Ah)   A, B   = constants
                             I.t   = actual battery charge (Ah)

Figure 7.6 Generic battery EMT model

in which an ideal controlled voltage source in series with a resistance is used to model the battery. At every time step the voltage of the controlled voltage source is computed based on the state of charge of the battery using two different methods. The first method is based on a nonlinear equation that uses the actual state of the battery to calculate the no-load voltage, and the value of the resistance is assumed to be constant. The second method of battery modeling uses characteristic curves of the battery to compute the value of the controlled voltage source and internal resistance based on the actual state of charge of the battery. In this method, SOC-voltage and SOC-resistance curves are required to model the battery.

8 Case Studies

The following three case studies illustrate practical aspects of planning energy storage assets. The author has conducted, with his colleagues at Quanta Technology Ltd., the analysis contained in each of the case studies in the course of their consulting practice in support of T&D utility customers. The information in each of the three case studies was published by the respective utility customer.

The name and application of each case study are summarized in Table 8.1.

Table 8.1 Summary of case studies

#	Case Study	Application
1	ATC – Waupaca Area Project	Transmission Grid Reliability (138 kV)
2	PSE – Bainbridge Island	Distribution Reliability (12 kV)
3	ISA – Integrating Wind in Colombia	Frequency Response and Low Inertia (110 kV)

8.1 Case Study 1: American Transmission Company (ATC) – Waupaca Project[3]

8.1.1 System Description and Problem Statement

The Waupaca area (Figure 8.1) of the ATC service territory in Wisconsin is supplied through two 69-kV sources (Wautoma to the south and Whitcomb to the north), one 115-kV source (Hoover to the west), and one 138-kV source (White Lake to the east). The peak load in the area reaches 200 MW and exceeds 155 MW during 6% of the days.

The area can potentially experience reliability concerns that involve thermal overloads and low-voltage conditions under scenarios involving multiple outages (i.e., NERC Category P6 or maintenance + P1 outages). Depending on the load level in the Waupaca area, thermal overloads, emergency low voltages, or unsolved contingencies may occur for the loss of a second line (i.e., loss of the remaining 115-kV or 138-kV source in the area). The most constraining outage in the area was observed to be the loss (planned or forced) of the 115-kV source at Hoover and the 138-kV source at White Lake. Under these multiple outages, the local loads cannot be sustained if the area load exceeds 120 MW.

The existing solution is to utilize an operating guide after the first outage to sectionalize the 69-kV system at certain load levels, creating radially served loads. This allows the loads to be served after the second contingency but places many loads at risk of loss for a subsequent single failure. This operating guide reduces maintenance opportunities and increases the amount of load at risk in the area.

The conventional wire solution consists of rebuilding the 115-kV source line as a double circuit, upgrading the 69-kV source bus to the south, and adding a 10-MVAR shunt capacitor at the Arnold 138-kV substation. The estimated

---

[3] https://cdn.misoenergy.org/20190531%20WSPM%20Item%2003d%20ATC%20SATOA%20review%20results349901.pdf

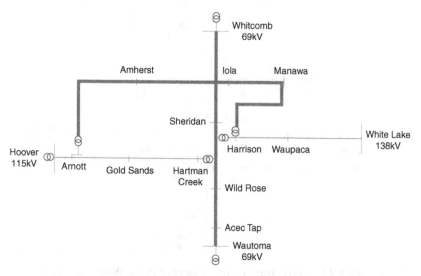

Figure 8.1 Waupaca region map

total capital cost of the conventional solution is $12.4 million (in 2019 dollars). This wire solution will require an expansion of the right-of-way (ROW).

ATC proposed a hybrid storage solution that aims to increase local area reliability and to provide operational flexibility in the Waupaca area. The hybrid storage solution consists of a 2.5 MW/5 MWh battery to be installed at the Harrison North 138-kV substation, 14 MVAR shunt capacitors (8 at Arnott and 6 at Harrison North), and an upgrade to the 69-kV source bus in the south. The estimated total capital cost of the solution is $9.1 M (in 2019$). This hybrid solution will have fewer public impacts on the right-of-way (ROW).

8.1.2 Planning Criteria

- Five-year planning horizon.
- Solution should address system reliability issues up to a peak load of 155 MW.
- Solution should, at a minimum, provide a two-hour window after the occurrence of the second contingency to reconfigure the 69-kV system without loss of load.
- Study Criteria: NERC TPL-001–4 and Applicable Local TO Planning Criteria

8.1.3 Storage Siting and Sizing Analysis

The ability to relieve the thermal and voltage violations through real and reactive power injections respectively is assessed at each bus in the study area (Figure 8-2). The minimum reactive power injection to relieve voltage

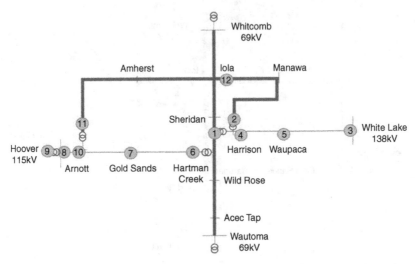

Figure 8.2 Potential sites of energy storage in the Waupaca region

violations and the minimum real power injection to relieve thermal overloads from each candidate bus are quantified and compared. Sites with the least amount of power injections are deemed more suitable than others. Additionally, the sites are further analyzed based on their topology to remain networked and not be isolated following the first critical contingency. A power flow program is used to calculate the minimum reactive power injection to maintain system voltages above 0.9 p.u. following the first contingency, and distribution factors are utilized to calculate the minimum MW injections to mitigate thermal overloads.

The overload on the Wautoma-ACEC Wautoma Tap can be mitigated by upgrading the bus rating to 94 MVA emergency rating and was excluded from the storage siting and sizing analysis. Table 8-2 presents the level of reactive and real power injections that are required to mitigate overloads and voltage violations if the solution is placed at one of the system buses.

The proposed hybrid project will mitigate the thermal overload and voltage violations in the system.

8.1.4 Battery Operation Modes

There are three operating modes that should be recognized and parametrized in the battery control system:

1. In the base (or intact) system, the battery will either be off-line during periods of the year when the forecasted load level does not trigger a reliability concern, or it can be used to regulate voltage using its dynamic reactive capability.

Table 8.2 Real and reactive power requirements

Map #	Bus Station Name	MVAR Need	MW Need Harrison Tap – Wild Rose Tap 69 kV	MW Need Harrison – Harrison Tap 69 kV
1	Harrison 69 kV	16	5.5	3.4
2	Harrison North 69 kV	16	5.5	3.4
3	White Lake 138 kV	16	5.9	3.6
4	Harrison 138 kV	16	5.9	3.6
5	Waupaca 138 kV	16	5.9	3.6
6	Hartman Creek 138 kV	16	6	3.6
7	Golden Sands 138	16	6.2	3.7
8	Hoover 138 kV	16	6.3	3.8
9	Hoover 115 kV	16	6.3	3.8
10	Arnott 138 kV	16	6.3	3.8
11	Arnott 69 kV	17	7.1	4.3
12	Iola 69 kV	19	8.4	5.1

2. After the first contingency in the target area, the shunt capacitors will be switched on, and the battery will be online but without injecting any real power. It can, however, regulate voltage if needed.

3. Following the second outage, the battery will automatically sense the line power flow and bus voltage and will discharge real power and regulate voltage to mitigate the thermal overloads and voltage violations.

Table 8.3 Comparative analysis of hybrid solution against conventional solution

	Wire Solution	Non-Wire Hybrid Solution
Solution Considered	Rebuild Whiting Avenue – Hoover 115 kV as a double circuit, install 10-MVAR capacitor at Arnott 138-kV substation, and upgrade Wautoma 69-kV bus	Install a 2.5-MW/5-MWh battery at Harrison North 138-kV substation, an 8-MVAR capacitor at Arnott 138-kV substation, a 6-MVAR capacitor at Harrison North 138-kV substation, and upgrade Wautoma 69-kV bus
Estimated Capital Cost (2019 $)	$11.3 M	$8.1 M
Present Value of Revenue Requirements (PVRR) for 40-Year Lifecycle Costs	$13.07 M	$12.24 M
Overall Comparison	Comparable performance More expensive Need for expanded ROW No online time restrictions	Comparable performance Less expensive No public impacts on ROW 2-hour discharge period

8.1.5 Comparative Techno-Economic Analysis

The relative economics of the non-wire hybrid solution (Table 8-3) was evaluated against the conventional wire solution. Project revenue requirements were calculated considering traditional utility rate-base accounting, including cost of capital, system installed costs, project management costs, substation adaptation costs, ongoing preventive and corrective maintenance, battery midlife capacity augmentations, inverter replacements, battery end-of-life disposal costs, and battery replacements cost; these costs were then discounted at the prevailing customer discount rate. The economic comparison was conducted over a 40-year period, comparable to the book life of a conventional transmission asset.

It is important to note that care should be exercised when comparing storage solutions to conventional T&D solutions. Each solution has additional attributes (e.g., benefits and risks) that have not been evaluated in this analysis. This analysis focused on the primary function of grid reliability up to a loading level of 155 MW. The economics and ease of expanding each solution to accommodate further load developments might be significantly different. Additionally, each type of solution might provide additional system, customer, or economic benefits that are not captured in this analysis. The study examined the potential revenue-stacking opportunities of the storage-only and the hybrid solutions and showed the possibility to lower the customer costs significantly. However, the uncertainty surrounding the ability to participate in the ancillary services and capacity markets, as well as the cost of market participation and market prices, should be carefully evaluated.

8.2 Case Study 2: Puget Sound Energy (PSE) – Bainbridge Island[4]

8.2.1 System Description and Problem Statement

The Bainbridge Island electric system needs assessment determined the island's grid has reliability, capacity, and aging infrastructure needs during the 10-year planning horizon on both the transmission and distribution systems. The Island's winter peak load is projected to grow at a compound annual growth rate (CAGR) of 1.23%, from 80 MVA in 2017 to 101 MVA in 2037, after accounting for demand-side management measures.

The Island is supplied by two 115-kV transmission lines from the Kitsap peninsula (Figure 8.3), and the loads are served from three 115/12.47-kV substations (Port Madison, Murden Cove, and Winslow) through four 600-amp feeders from each substation (Figure 8.4). Each substation has one transformer with a winter normal rating of 33 MVA and an emergency rating of 36 MVA. Reclosers allow the transfer of load between feeders following outage events.

System needs and concerns for Bainbridge Island are as follows:

1. **Transmission Reliability Need**: A reliability improvement need was identified to improve the performance of the Winslow Tap transmission line that feeds Winslow substation. The Winslow substation experienced 22 outages

---

[4] https://oohpsebainbridgefall2019.blob.core.windows.net/media/Default/documents/Appendix%20E_Energy%20Storage%20Planning%20to%20Support%20Bainbridge%20Island%20Final%20Report_Quanta%20Technology_Apr%2023%202019%204a.pdf

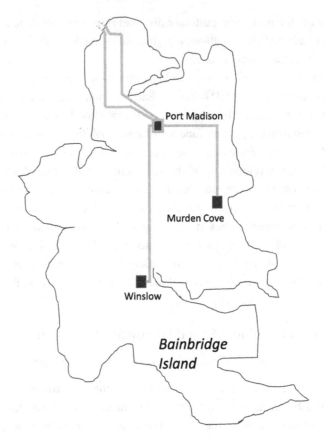

Figure 8.3 Bainbridge Island study area

over a six-year period (2012–2017), an average of nearly four substation outages per year. Nearly 70% (15 out of 22) of the Winslow substation outages were caused by the loss of Winslow transmission tap due to tree-related events. Therefore, the storage solution should be sized to securely serve the load that would have been interrupted after the outage of the Winslow substation, while accounting for any potential support from the Murden Cove and Port Madison substations.

2. **Substation Capacity Need**: Considering PSE planning guidelines, the substation group capacity planning trigger of 85% (or 84 MVA) for the Winslow, Murden Cove, and Port Madison substations will be exceeded after the addition of a Ferry charging station load connected to the Murden Cove distribution station. A distribution substation group capacity need of 14.6 MW was identified on Bainbridge Island within the 10-year planning horizon (2018–2027) to support general load growth of 4.6 MW and planned

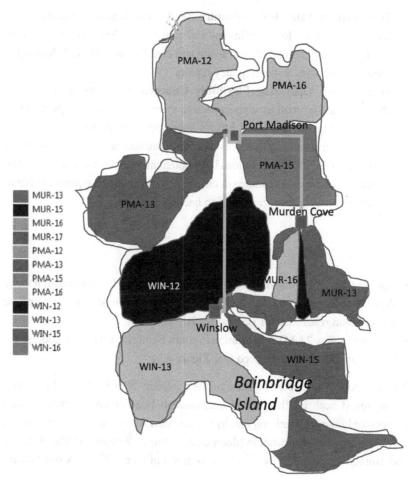

Figure 8.4 Distribution feeders on Bainbridge Island

10-MW load addition for a planned ferry charging station load. The anticipated capacity need is expected to grow to 16.6 MW by 2030 due to general
load growth increase by 2 MW. Per the PSE Solution criteria, a solution must
last 10 years. The Needs Assessment shows that additional substation
capacity is needed by 2020. Therefore, the need of 16.6 MW is the ultimate
need for a viable solution to last until 2030.

3. **Distribution Feeder Reliability Need**: The storage solution should be sized
   to carry the entire load served by the Winslow-13 feeder until the restoration
   work is completed (estimated at four hours), to improve the reliability
   indices of system average interruption duration index (SAIDI) and customer
   minutes of interruption (CMI).

4. **Transmission Aging Infrastructure Need**: An infrastructure replacement need was identified for the Winslow Tap transmission line support structures that are nearing the end of useful life and could potentially fail, leading to unplanned outages and emergency repairs.
5. **Transmission Operating Flexibility Concern**: Concerns related to the ability to transfer load to support routine maintenance and outage management. The Winslow and Murden Cove substations are on radial transmission taps (single transmission source) with no transmission backup. Customers served from these two substations have potential for outage in the event of an unplanned transmission outage or emergency transmission equipment repair situation due to lack of transmission backup.

8.2.2 Conventional Solution

- Construct a 115-kV transmission line from the Murden Cove substation to the Winslow substation to create a looped transmission system to improve transmission reliability for Bainbridge Island.
- Construct a new 25-MVA substation in south Bainbridge Island to address the substation group capacity need (see Figure 8.5).

This solution was selected because of the reliability benefit. Reliability would be improved both by the looped transmission line (decreased transmission outages and outage durations) and by the decrease in the number of customers served by any single substation (decreased customer outages) for the Winslow and Murden Cove substations with the addition of a new 25-MVA substation.

8.2.3 Energy Storage Solution – Summary

The study focused on a 10-year planning horizon (2018 through 2027) using a baseline load forecast scenario that includes Demand Side Management (DSM) and a Ferry charging station starting its operation in 2021. It addressed three system needs,[5] namely:

- Transmission reliability (Winslow Tap outages)[6]
- Substation group capacity (e.g., sum of all substations)
- Feeder reliability (Winslow-13)

---

[5] Based on Bainbridge Island Electric System Needs Assessment Report, PSE Strategic System Planning, May 14, 2018 draft.
[6] PSE required eight-hour backup of substation load for a transmission outage to provide sufficient time for crews to restore transmission service. A four-hour backup of feeder load was required for feeder outage restoration.

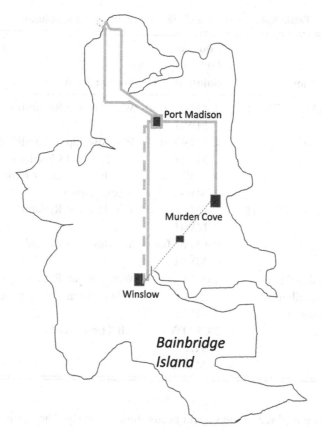

Figure 8.5 Proposed conventional solution

The analysis yielded a solution alternative consisting of five batteries with a total combined rating of 29.3 MW / 91.2 MWh. The batteries would be distributed in the island at locations (shown in Table 8.4 and Figure 8.6) along circuits or at a substation. One of the five batteries could be located at substation property, while the four other large batteries would be at locations determined around the island on specific circuits and not at a substation. The storage-only solution alternative would cost at least $20 million more than the conventional wire solution as documented further in the study; this alternative's cost will increase due to interconnection costs and siting of the batteries.

The storage solution's overall size was methodically optimized by exploiting the following levers: (1) existing PSE feeder switching schemes, (2) proposing modifications to the switching schemes to enable shifting more loads between feeders, and (3) finding sites where one storage solution can

Table 8.4 Locations and sizes of energy storage solution

ID	Location	Storage-Only Solution	System Need
1	PMA-13/WIN-12	3.2 MW/ 9 MWH	Winslow Tap Reliability
2	WIN-13	4.4 MW/20 MWH	Winslow Tap Reliability & Winslow-13 Feeder Reliability
		4.2 MW/12 MWH	Winslow-13 Feeder Reliability (exclusive)
3	MUR-17/WIN-15	3.4 MW/ 15 MWH	Winslow Tap Reliability
4	MUR-15	0.4 MW/ 0.4 MWH	Winslow Tap Reliability
5	Murden Cove Distribution Station	13.7 MW/ 34.8 MWH	Winslow Tap Reliability & Substation Capacity Needs
	Total	**29.3 MW / 91.2 MWH**	**All Three Needs**

address two or three of the system needs simultaneously. The detailed siting and sizing analysis section illustrates all the steps taken in this study to optimize the storage solution.

8.2.4 Storage Siting and Sizing Analysis Methodology

The storage systems are initially sited and sized to address each of the three system constraints, and then optimized to leverage their locational synergies to address multiple (i.e., two or three) system needs simultaneously. The following elaborates on the methodology used to address each system need.

8.2.4.1 Storage Siting and Sizing for Winslow Tap Reliability

- Storage size (MW and MWh) is optimized to provide backup to Winslow substation load for up to eight hours[7] after outage of the Winslow transmission tap.

---

[7] PSE required an eight-hour backup of substation load for a transmission outage to provide sufficient time for crews to restore transmission service. A four-hour backup of feeder load was required for feeder outage restoration.

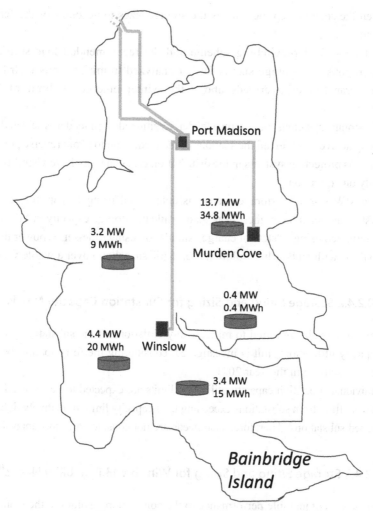

Figure 8.6 Energy storage locations

- Using PSE's existing switching schemes, the feeders at the Winslow substation are first switched onto feeders at the Murden Cove and Port Madison distribution stations and then the storage size requirements for each overloaded feeder are determined.
- The possibility of load shifting from the heavily loaded to the lightly loaded feeders is investigated to optimize the storage size requirements.
- Due to the nature of the load profile of the ferry charging station served by the Murden Cove substation, which differs from the other system load profiles, the analysis of the storage sizes is taken in two steps. The initial step of the analysis quantifies the storage sizes without the ferry load, and

then the second step increments the storage sizes to account for the ferry load.

- After the feeder-switching scheme and the recommended load-shifting operations, the storage size is further analyzed to mitigate any distribution transformer overloads above the winter emergency limit of 36 MVA.
- If a storage system is located on a feeder, its siting along a feeder is selected to mitigate overloads on all the sections of the feeder and to avoid reverse power flow, as protection systems in the distribution grid generally have visibility in only one direction.
- The MW size of a storage system is determined using a snapshot of the system model at the highest peak load, while the storage capacity in MWh is determined using a "state of charge" simulation using time-series power flow analysis with an hourly resolution (i.e., 8760 snapshots) over a whole year.

8.2.4.2 Storage Siting and Sizing for Substation Capacity Needs

- The storage size required to mitigate any violations of the substation group capacity utilization limit is investigated, considering the ferry load to be in service starting in the year 2021.
- Individual substation capacity utilization limits are expected to be resolved by load shifting from substations exceeding their capacity limit to relatively lightly loaded substations. Therefore, such needs are not considered in this analysis.

8.2.4.3 Storage Siting and Sizing for Winslow 13 Reliability Needs[8]

- To have a comparable performance to the conventional solution, the storage size is optimized to back up the entire feeder's load for four hours.

A view of the detailed sizing and placement analysis is summarized in Table 8.5. Each of the three system needs is addressed separately while the storage systems are sized and placed to serve multiple system needs, thus optimizing the overall storage system size and cost.

8.2.5 Techno-Economic Comparative Analysis

The capital cost of the storage solution is compared to the cost of the conventional solution in Table 8-6 and shows the following:

---

[8] Bainbridge Island Electric System Needs Assessment, PSE Strategic System Planning, May 14, 2018.

Table 8.5 Summary of battery solution siting and sizing analysis

		Battery Size		Needs →				Overloads →			
ID	Site Location	Size MW	Size MWh	WIN-TAP	Sub. Group	WIN-13 Feeder	MUR-16	MUR-17	PMA-13	MC Trafo	Placement Options
1	PMA-13 / WIN-12	3.2	9	X					X		1.2–2.8 miles from PMA sub, along PMA-13 or WIN-12
2	WIN-13	4.4	20	X		X	X			X	1.0–3.5 miles from MC sub, along MUR-16, WIN-16, or WIN-13
	WIN-13	4.2	12			X					WIN-13
3	MUR-17/ WIN-15	3.4	15	X				X		X	2.2–4.8 miles from MC sub, along MUR-17 or WIN-15
4	MUR-15	0.4	0.4	X				X		X	1.0–2.9 miles from MC sub, along MUR-15
5	MC Sub	13.7	34.8	X	X					X	MC sub
	Total Battery Size	29.3	91.2	25.1 MW/ 79.2 MWh	Included in WIN-TAP	8.6 MW/ 32 MWh	4.4 MW/ 20 MWh	3.8 MW/ 15.4 MWh	3.2 MW/ 9 MWh	21.9 MW/ 70.2 MWh	Note: MC is Murden Cove Substation

Table 8.6 Comparison between conventional and storage solution initial costs

| Need Driver | Need Year | Conventional T&D | | ALL-BESS Option | |
		Solution	Costs[*]	Storage Sizes (MW/MWH)	Costs
Transmission Reliability – Winslow	Current	Transmission Loop	$12,300,000	25.1 MW/ 79.2 MWH	$31,923,804
Substation Group N-0 Capacity	2021	New Distribution Substation	$11,250,000	9.7 MW/ 5 MWH	$4,077,290
Feeder Reliability (WIN-13)	Current	Conventional feeder reliability solution, $640 k underground conversion	$640,000	8.6 MW/ 32 MWH	$12,550,531
ALL		ALL the above	$24,190,000[a]	29.3 MW/ 91.2 MWh	$36,909,791
Upsizing for Degradation				29.3 MW / 111 MWh	$43,500,000

[*] Costs are from a July 2018 Puget Sound Energy cost estimate based on similar past projects in other areas of PSE service territory. Does not include site-specific engineering.

- The size of the storage system to address all three system needs is less than the sum of storage system sizes to address each of the three individual system needs. This is a result of properly siting the storage to address multiple needs simultaneously wherever applicable.
- The capital cost of the storage solution to address all three system needs is estimated to be 80% higher than the cost of the conventional solution.
- The storage solution is only competitive to address one of three system needs, namely the Substation Group limit.

The lifetime economic analysis of the optimized storage solutions was compared against the conventional T&D solution over the 10-year planning horizon, and the results are tabulated in Table 8-7.

The analysis included all the relevant cost components such as capital and O&M, considering storage lifecycles, efficiency, and capacity fading issues. However, it does not include interconnection costs, land and permitting costs, and other costs associated with distribution automation. This study finds the storage-only solution to be over three times more expensive in meeting the system needs than the conventional T&D solution.

The economic evaluation of the storage solutions as compared to the conventional T&D solutions requires:

- Lifetime modeling of the cost of each project from inception to retirement inclusive of project development activities and timeline, EPC, O&M, capacity management, replacement, and disassembly and recycling.
- Modeling of relevant utility's capital structure, including debt and equity ratios and costs, and tax rate.
- Proper regulated asset base (RAB) accounting, including treatment of depreciation for tax and book purposes.
- Useful life estimates: The conventional T&D solutions have an assumed book life of 45 years, while the energy storage technology is assumed to have a useful life of 15 years for Li-ion technology (20 years for Flow technology).

This study adopted the following methodology to compare the economics of the various solution alternatives:

1. The capacity of the energy storage system is upsized to mitigate the anticipated degradation over the 10-year planning horizon. For a nominal 2% annual degradation in storage capacity, the storage MWh capacity is upsized by 22% from the level required to address the system reliability needs.
2. The capital cost components of each solution alternative are calculated (conventional T&D, and energy storage).

Table 8.7 Comparison between conventional and storage solution lifetime costs

All Costs Are Present Value ($M)	Conventional T&D Solution[*]	Storage-Only Solution[**]	Storage-Only Solution (Option)
Application	Distribution Capacity & Reliability	Distribution Capacity & Reliability	Distribution Capacity & Reliability (excluding WIN-13 feeder reliability)
Project Need Date	2018	2018	2018
Storage Size MW/MWh			
Min Size to Meet System Needs	-	29.3 MW / 91.2 MWh	25.1 MW / 79.2 MWh
Upsized to Mitigate Degradation		29.3 MW / 111 MWh	25.1 MW / 97 MWh
Capital Investment –Conventional Storage	$24.2[†]	$43.2[††]	$37.7[††]
Total	**$24.2[†]**	**$43.5[††]**	**$37.7[††]**
Capital Levelized Real Cost[†††] (over 10 years)	$10.0	$32.6	$28.2
O&M Cost (over 10 years)	$0.4	$1.6	$1.4
Total Cost (over 10 years)	**$10.4**	**$34.1**	**$29.6**
Cost Ratio	**100%**	**328%**	**284%**

* Conventional T&D solution asset life of 45 years.

** Storage-only solution asset life of 15 years.

† Costs are July 2018 PSE cost estimate based on similar past projects in other areas of PSE service territory. Does not include site-specific engineering.

†† Costs do not include interconnection costs, land and permitting costs, and other costs associated with distribution automation.

3. However, because of the differences in asset life between the conventional component (45 years) and the storage component (15 years) of any solution, the cost of each component over the 10-year planning horizon is calculated and summed (using present value) to provide a total 10-year capital cost. This calculation utilizes a real economic carrying cost, considering the company's weighted average cost of capital and the inflation rate.
4. The present value of the O&M costs over the 10-year planning horizon is calculated and summed for each solution.
5. The overall (capital and O&M) present value costs of all solutions are calculated and compared.

It is important to note that care should be exercised when comparing storage solutions to conventional T&D solutions. Each solution has additional attributes (e.g., benefits and risks) that have not been evaluated in this analysis. For example, the energy storage solution can address reliability needs for a finite amount of time (assuming eight hours for transmission outages and four hours for distribution outages), whereas the conventional T&D solution provides a solution with an indefinite time. This analysis focused on the primary function of each solution in terms of grid capacity and/or reliability to accommodate projected load development for a period of 10 years beyond the installation date (2019–29). It is challenging to procure and install the battery systems within 12 months, although it is not impossible. Beyond that period, the economics and ease of expanding each solution to accommodate further load development might be significantly different. Additionally, each type of solution might provide additional system, customer, or economic benefits that are not captured in this analysis.

8.2.6 Economic Analysis Results with Stacked Market Revenues

The study examined in detail the economic potential of the storage assets to provide additional services beyond the three system needs. Two potential services that were available to the utility without an organized energy market were analyzed: the system capacity service and energy arbitrage.

The additional revenues generated by the excess capacity of the storage assets after meeting the reliability system needs were relatively small, with a 10-year present value of $2.4 million for the Storage-Only solution, and $2.1 million for the Storage-Only (Option) solution. Thus, after accounting for the revenue-stacking opportunities, the storage solutions were still around three times the cost of the conventional solution.

The Bainbridge Island system needs vary throughout the year with the load level, and correspondingly, the percentage of the battery size (MW and MWh) that is required to address these system needs will also vary. This provides a commercial opportunity to monetize the excess capacity of the storage solution during periods when the needs are not at their highest levels, to offset the cost of the storage investment.

Two monetizable services that are potentially available to this storage system are system capacity and energy price arbitrage. The Integrated Resource Plan (IRP) of 2017 provides a basis for the valuation of these two potential revenue streams. System capacity is a contracted service and thus has a lower risk as compared to energy price arbitrage, which depends on the daily volatility of the locational marginal prices (LMP). Due to the small size of the storage solution as compared to the overall utility load, the impact of the storage system operation on locational marginal prices is negligible, and thus the analysis will assume the storage to be a price taker and will not account for potential reduction of cost of energy to utility customers. The first priority of utilizing the storage system is to address the local needs. Any excess capacity will be utilized to provide system capacity services, and finally, any remaining excess capacity will be used to provide energy price arbitrage.

The analysis methodology that was followed in this study to quantify and optimize the revenue streams is summarized in the following three steps:

- Determine the required storage capacity (MW and MWh) for each hour in a year to address the three local needs.
- Assess the storage availability to provide the system capacity service.
- Optimize the operating profits from energy price arbitrage.

8.2.6.1 Hourly Storage Requirements to Address Needs

The storage solution consisting of five storage systems was designed to address three needs. Two of these needs (Winslow-Tap outage and WIN-13 feeder outage) are contingent services that are triggered only after the onset of a defined grid outage, while the third need (substation group) is triggered whenever the load level exceeds a prescribed level and thus is not contingent. To address the two types of needs, the storage systems will have to discharge to address the substation group trigger while keeping enough energy (MWh) in reserve to potentially address a subsequent defined outage. Both the discharged MW and the reserve MWh vary hourly throughout the year, depending on the load level on specific feeders and substations in the area.

Figure 8.7 shows the percentage of the storage solution size that is required to address the substation group capacity requirement, while Figure 8.8 shows the storage solution size (expressed as a percentage of the total planned solution size) that is required to address all three needs. Figure 8.9 shows the remaining

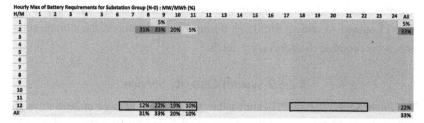

Hourly Max of Battery Requirements for Substation Group (N-0) : MW/MWh (%)

H/M	1	2	3	4	5	6	7	8	9	10	11	12	13	14	15	16	17	18	19	20	21	22	23	24	All
1									5%																5%
2								31%	33%	20%	5%														33%
3																									
4																									
5																									
6																									
7																									
8																									
9																									
10																									
11																									
12								12%	22%	19%	10%														22%
All								31%	33%	20%	10%														33%

Figure 8.7 Storage max hourly requirements for substationg capacity need by month and hour (taken as the higher of MW% or MWh%)

Hourly Max of Battery Requirements MW/MWh (%)

H/M	1	2	3	4	5	6	7	8	9	10	11	12	13	14	15	16	17	18	19	20	21	22	23	24	All
1	12%	17%	18%	20%	22%	30%	42%	47%	49%	45%	43%	39%	33%	26%	26%	26%	29%	38%	41%	41%	37%	33%	17%	11%	49%
2	26%	37%	49%	61%	69%	83%	84%	79%	69%	58%	51%	49%	47%	47%	72%	71%	66%	61%	63%	59%	53%	48%	36%	22%	84%
3	10%	11%	12%	13%	15%	16%	30%	44%	46%	44%	35%	30%	17%	15%	13%	13%	14%	17%	20%	23%	19%	15%	10%	10%	46%
4	9%	9%	10%	12%	14%	16%	27%	45%	44%	30%	15%	11%	10%	10%	10%	16%	11%	11%	11%	11%	10%	9%	9%	9%	45%
5	8%	8%	8%	10%	11%	13%	21%	46%	49%	44%	40%	30%	17%	9%	10%	14%	16%	18%	29%	29%	28%	23%	9%	8%	49%
6	7%	7%	7%	7%	7%	7%	8%	8%	8%	8%	7%	7%	7%	7%	7%	7%	7%	8%	8%	8%	7%	7%	7%	7%	8%
7	7%	7%	7%	7%	7%	7%	7%	7%	7%	7%	7%	7%	7%	7%	7%	7%	7%	7%	7%	7%	7%	7%	7%	7%	7%
8	6%	6%	6%	6%	6%	6%	6%	6%	6%	6%	6%	6%	6%	6%	6%	6%	6%	8%	8%	8%	7%	7%	6%	6%	8%
9	5%	5%	5%	5%	7%	8%	9%	9%	9%	8%	7%	7%	7%	7%	7%	7%	8%	8%	9%	9%	8%	7%	6%	5%	9%
10	7%	7%	7%	9%	10%	12%	13%	25%	33%	25%	10%	10%	10%	10%	10%	10%	11%	11%	11%	10%	9%	8%	7%	7%	33%
11	12%	14%	16%	18%	20%	23%	39%	47%	47%	46%	44%	42%	40%	40%	35%	35%	34%	41%	39%	38%	35%	32%	20%	12%	47%
12	19%	26%	33%	40%	46%	56%	65%	65%	65%	51%	55%	50%	47%	45%	42%	40%	41%	44%	45%	45%	44%	42%	37%	25%	65%
All	26%	37%	49%	61%	69%	83%	84%	79%	69%	58%	51%	49%	47%	47%	72%	71%	66%	61%	63%	59%	53%	48%	37%	25%	84%

Figure 8.8 Storage max hourly requirements for all three needs by month and hour (taken as the higher of MW% or MWh%)

	1	2	3	4	5	6	7	8	9	10	11	12	13	14	15	16	17	18	19	20	21	22	23	24
1	88%	83%	82%	80%	78%	70%	58%	53%	51%	55%	57%	61%	67%	74%	74%	74%	71%	62%	59%	59%	63%	67%	83%	89%
2	74%	63%	51%	39%	31%	17%	16%	21%	31%	42%	49%	51%	53%	53%	28%	29%	34%	39%	37%	41%	47%	52%	64%	78%
3	90%	89%	88%	87%	85%	84%	70%	56%	54%	56%	65%	70%	83%	85%	87%	87%	86%	83%	80%	77%	81%	85%	90%	90%
4	91%	91%	90%	88%	86%	84%	73%	55%	56%	70%	85%	89%	90%	90%	90%	84%	89%	89%	89%	89%	90%	91%	91%	91%
5	92%	92%	92%	90%	89%	87%	79%	54%	51%	56%	60%	70%	83%	91%	90%	86%	84%	82%	71%	71%	72%	77%	91%	92%
6	93%	93%	93%	93%	93%	93%	92%	92%	92%	92%	93%	93%	93%	93%	93%	93%	93%	92%	92%	92%	93%	93%	93%	93%
7	93%	93%	93%	93%	93%	93%	93%	93%	93%	93%	93%	93%	93%	93%	93%	93%	93%	93%	93%	93%	93%	93%	93%	93%
8	94%	94%	94%	94%	94%	94%	94%	94%	94%	94%	94%	94%	94%	94%	94%	94%	93%	92%	92%	92%	93%	93%	94%	94%
9	95%	95%	95%	95%	93%	92%	91%	91%	91%	92%	92%	93%	93%	93%	93%	92%	92%	92%	91%	91%	92%	93%	94%	95%
10	93%	93%	93%	91%	90%	88%	87%	75%	67%	75%	90%	90%	90%	90%	90%	90%	89%	89%	89%	90%	90%	91%	92%	93%
11	88%	86%	84%	82%	80%	77%	61%	53%	53%	54%	56%	58%	60%	60%	65%	65%	66%	59%	61%	62%	65%	68%	80%	88%
12	81%	74%	67%	60%	54%	44%	35%	35%	39%	45%	50%	53%	55%	58%	60%	59%	57%	56%	55%	55%	56%	58%	63%	75%

Figure 8.9 Available storage capacity after meeting system needs

storage capacity after addressing all reliability needs. The percentage size is conservatively computed as the higher of either MW rating percentage or MWh rating percentage. The analysis was conducted using the 2019 data, which corresponds to the highest peak in the study horizon. The data is tabulated in columns corresponding to each of the 24 hours in a day, and in rows corresponding to each of the 12 months in a year. For example, at hour ending 8, the maximum discharge for all days in December to address the substation group capacity need will be 22% of the storage solution rating, and at the same time, the storage solution will have to maintain 65% of its rated energy capacity in reserve in anticipation of an outage in accordance with the Winslow-Tap or the Win-13 outage scenarios. Similarly, during June–September, the storage solution is not required to address any needs.

8.2.6.2 System Capacity Service

The Puget Sound IRP in 2017 shows the system capacity price in 2018 through 2024 to be $3.79/kW-yr, and then jumps in 2025 to $78.19/kW-yr and stays at that level through 2037. The system capacity requirements are exclusively during December for each of the 10 hours between 6 AM and 11 AM and between 5 PM and 10 PM.

An analysis of the local requirements in Figure 8.7 and Figure 8.8 and the hourly profile of the load in Bainbridge Island reveals the following observations:

- The load peaks in December during hour-ending HE 8–11 AM.
- The energy required to shave the peak load in December increases with the level of MW shaving. A 5 MW peak load shaving requires three hours of energy, while a 10 MW peak shaving requires four hours, and further peak shaving beyond 15 MW requires 10 hours.
- During December, the maximum MW discharge of the storage solution during the system capacity hours is 22% of the storage solution rating, while the maximum energy requirement is 65% of the storage rating. At any hour, the MW discharge in addition to the excess storage capacity above the requirement is available to provide the system capacity service. The minimum available percentage of the storage rating for system capacity services is 35%.
- During the 10 system capacity hours in December, the availability of the storage solution to provide system capacity service rises, as a percent of the storage solution rating, from 35% at 7 AM to 58% at 10 PM.
- The storage solution rating of 29.3 MW and 91.2 MWh provides only 3.11 hours of continuous discharge capability.
- If the storage aims to provide an equal level of system capacity at each of the 10 hours for each of the days in December, then the maximum level of participation in the system capacity service will be (3.11 hr/10 hr) × (35%) × (29.3 MW) = 3.2 MW, earning the full system capacity price. The present value of the system capacity service over 10 years is $0.63 million.
- On the other hand, if the storage solution is viewed as a component of a portfolio of capacity solutions and thus is allowed to provide partial capacity for each hour, the storage can maximize its participation by focusing on the last three hours in the day where it can participate by an average of 70%

of its rating, or $(56\%) \times (29.3) = 16.4$ MW, earning only 30% of the system capacity value (due to its limited energy capacity of three hrs). The present value of providing system capacity over 10 years is $0.86 million.

- Taking the average of the above two methods provides an estimated capacity value of $0.75 million.

8.2.6.3 Energy Price Arbitrage

During the hours when the storage is not required for reliability or not providing system capacity service, it has the potential to arbitrage the energy price by charging during periods of low prices and discharging during periods of high energy prices. The arbitrage potential is analyzed on a day-by-day basis, and only once a day at most to avoid excessive utilization of the storage life cycles. For a storage system with three hours of energy capacity, the first storage hour has the potential to generate the most profits because it can discharge at the highest price hour in a day and charge at the lowest price hour in the same day. The second hour of energy capacity will have to settle for the second-best discharge and charge hours and, thus, generate less revenue. Adding all the arbitrage profits after deducting the cost of round-trip losses provides an estimate of the maximum potential revenue from participation in this service.

The hourly locational energy prices (provided by Puget Sound according to the 2017 IRP) between 2018 and 2037 are averaged and summarized in Figure 8.10 by month and hour of the day and color coded, with red being highest and green being lowest. The average LMP is $56/MWh across the period with a high of $155/MWh and a low of $14/MWh. The average daily profile of the hourly prices is displayed in Figure 8.11 for five individual years between 2019 and 2037, along with the average over all the years. The daily price profile shows a rising LMP level year-over-year and a daily peak around hour-ending 19.

Average LMP ($/MWh)

M/H	1	2	3	4	5	6	7	8	9	10	11	12	13	14	15	16	17	18	19	20	21	22	23	24	Monthly
1	53	51	50	50	50	53	58	59	58	57	56	56	56	55	55	55	57	63	67	67	65	62	59	56	57.0
2	54	52	50	50	51	53	57	58	58	58	57	57	55	55	55	56	61	66	68	67	64	61	57		57.3
3	50	49	49	49	49	52	54	55	55	56	55	55	54	53	52	52	53	55	60	62	62	60	55	51	54.1
4	48	47	47	47	47	48	49	50	52	52	52	51	51	50	49	49	50	53	58	59	60	58	53	50	51.2
5	46	45	45	45	46	45	45	45	45	46	46	47	47	47	48	48	50	53	56	56	55	53	49	47	48.1
6	45	45	45	45	46	45	45	45	45	47	47	48	49	50	52	54	56	57	57	55	53	50	49	46	49.1
7	49	47	47	47	47	47	47	47	48	50	51	54	56	58	61	64	69	71	70	65	60	58	56	52	55.1
8	55	52	51	51	50	51	50	50	51	52	54	57	60	62	66	73	79	81	81	75	69	64	60	58	60.4
9	57	55	53	52	53	53	55	54	56	57	58	59	60	61	64	68	73	79	80	77	72	65	62	60	61.9
10	54	53	52	52	53	56	58	58	58	59	60	60	60	60	61	64	69	71	71	68	64	61	57		59.9
11	56	54	53	53	53	56	59	58	58	59	59	58	58	59	62	66	68	69	67	64	62	58			59.4
12	57	54	53	53	53	55	59	58	58	57	57	57	57	57	58	61	67	70	69	67	65	62	59		59.2
AVG	52	50	50	49	50	51	53	53	53	54	55	55	55	56	56	58	61	65	67	66	64	61	57	54	56
Min	14	14	14	14	15	15	16	18	20	20	20	20	21	21	21	22	21	20	20	20	20	20	16	16	14
Max	97	99	95	95	95	100	108	108	103	100	100	99	105	108	121	131	142	155	155	145	133	113	108	101	155

Figure 8.10 Average Locational Marginal Prices (LMP) by month and hour

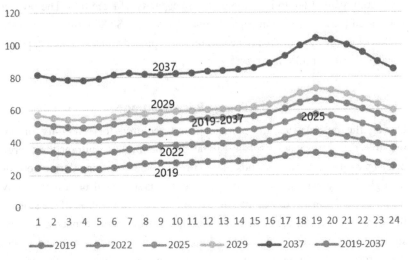

Figure 8.11 Average daily profile (24 hours)

The daily arbitrage profit potential for each of the storage capacity hours is shown in Figure 8.12 in cumulative format for an average year between 2018 and 2037. The maximum arbitrage gross profit potential of the first storage capacity hour reaches as high as $70/MWh and can only be profitable during 290 days of the year, while the gross profit of the fourth storage capacity hour reaches a high of $50/MWh and stays profitable for only 200 days in a year. The annual potential profit from participating in energy arbitrage is shown in Figure 8.13. The maximum profit potential is $3.8/kW-yr for the first storage hour and drops to $1.9 for the fourth hour. The rise in arbitrage profits over time is shown in Figure 8.14 as the LMPs rise. The profits are organized into two groups; one when the storage is allowed to participate every day in the year, while the second group restricts participation to only 200 days per year.

The Storage-Only solution, being a three-hour battery, has the potential to earn $6.6/kW-yr ($2.6 + $2.2 + $1.8) in 2019, which increases to $10.6/kW-yr in 15 years. This totals an average of $252,000 annually and a present value of $1.66 M over 10 years.

8.2.6.4 Optimized Revenue Stacking

Beyond the reliability and capacity requirements, the analysis reveals that the storage solution has the potential to earn $2.4 million, at most, in additional

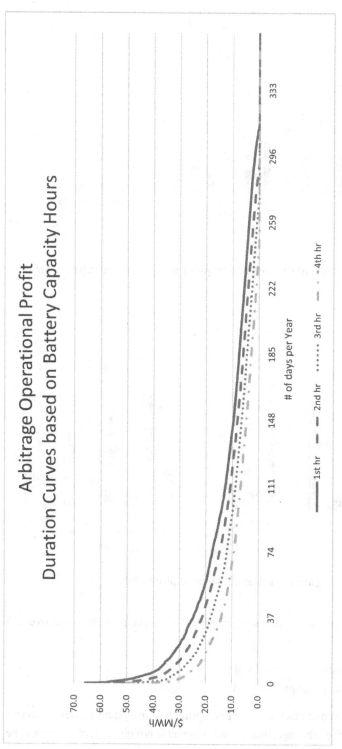

Figure 8.12 Cumulative arbitrage gross profit for 1–4 storage capacity hours

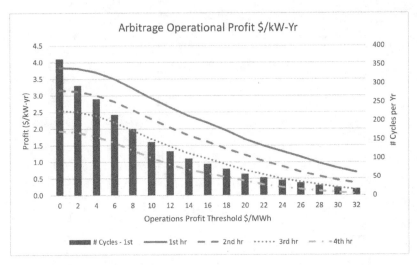

Figure 8.13 Arbitrage gross annual profit for 1–4 storage capacity hours

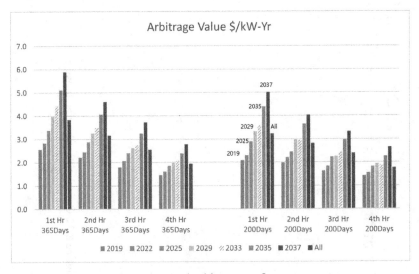

Figure 8.14 Annual arbitrage profit (2019–2037)

revenues when optimized using historical data (in present value over 15 years):

- System capacity $0.75 M
- Energy price arbitrage $1.66 M

Even if the revenue stacking could be optimized for system capacity services and energy price arbitrage, this revenue amount is not guaranteed and would be dependent on many other factors including perfect knowledge of market forward price curves, perfect equipment performance, and wholesale price growth in line with the IRP assumption.

8.3 Case Study 3: Wind Integration in the Caribbean Region of Colombia

8.3.1 Background and Application Definition

Several developers have proposed the construction of up to 3 GW of wind resources in the Guajira zone of the Caribbean region in Colombia. UPME (Energy Mining Planning Unit) has developed several transmission upgrade options to integrate these resources, including HVAC and HVDC lines. UPME also simulated the frequency response of the Colombian power system to assess the impact of low inertia and the ability to operate within the safe reliability limits without triggering under-frequency load shedding schemes (UFLS). The simulation results show a reliability violation in response to the loss of a 400-MW plant as the frequency drops below 59.5 Hz.

The Caribbean region differs from the central regions of Colombia by its reliance on mostly thermal plants as opposed to hydro plants. The grid is also limited and congested, especially in the 110-kV network. The region is interconnected with the central regions through 3 × 500-kV lines with a capacity of 1,500 MW. See Figure 8.15.

Wind integration can pose some challenges to system operations. A catalog of challenges and potential solutions is summarized in Table 8.8.

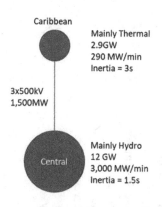

Dispatchable	Caribbean	Central	Total MW	% Total
Hydro	338	10,554	10,892	3%
APCM	473	774	1247	38%
Coal	464	879	1343	35%
Gas	1331	197	1528	87%
Fuel Oil	299	0	299	100%
Jet Oil/Gas	0	310	310	0%
Subtotal	**2,905**	**12,714**	**15,619**	**19%**

Non-Dispatchable	Caribbean	Central	Total MW	% Total
Co-Gen	0	84	84	0%
Small Hydro	0	661	661	0%
Wind	18	0	18	100%
Small Gas	31	90	121	26%
Subtotal	**49**	**751**	**800**	**6%**

| Grand Total | 2,954 | 13,465 | 16,419 | 18% |

Caribbean
Mainly Thermal
2.9GW
290 MW/min
Inertia = 3s

3x500kV
1,500MW

Central
Mainly Hydro
12 GW
3,000 MW/min
Inertia = 1.5s

Figure 8.15 Caribbean and central power system characteristics in Colombia

Table 8-8 List of possible challenges (columns) and potential solutions (rows)

Challenges/ Remedies	Production Predictability	Alignment with Load Profile	Short-Term Power Ramps	Reduction in System Inertia	Reduction in Fault Currents	Interconnection Grid Capacity
System Size Limits				✓		
Geographic Disbursement			✓			
Production Curtailment			✓			✓
Smart Inverters						✓
Forecasting	✓		✓			
Increase System Ramp Rates			✓	✓		
Decrease System Pmin		✓				
Flexible Reserves	✓					
Battery – Power			✓	✓		
Battery – Energy		✓				✓ (N-1-x)

Grid Expansion

Expand the Market Footprint

Shorten the Market Cycle

Innovative Protection Scheme

Several analyses were conducted to assess the challenges of integrating up to 3 GW of wind plants, including wind resources assessment, curtailment analysis, power ramp rate assessment, and system frequency response analysis. The results of these analyses are described next.

8.3.2 System Frequency Response Analysis

UPME simulated the loss of 400-MW plant under two wind penetration scenarios: 1.4 GW and 2.9 GW. The system frequency drops at a rate of 1.7 Hz/s in the 1.4-GW scenario, and at 2.0 Hz/s in the 2.9-GW scenario. This rate of change of frequency (RoCoF) does not pose a risk to the safety of equipment. However, the analysis showed the system frequency response to dip below 59.5 Hz level, thus violating the safe operating limits and possibly triggering underfrequency load-shedding relays.

The 400-MW contingency caused the system frequency to drop below 59.4 Hz within three seconds. Trying to counteract this large contingency using the primary frequency response of additional generation units will be difficult and will require adding a large fast-ramping generator to the system such as a diesel combustion engine or an aero-derivative turbine. Alternatively, an energy storage system can rapidly and precisely counteract the sudden drop in frequency.

To assess the required size of energy storage, a simplified model was created to mimic the system frequency response. Subsequently, a battery storage system was added to demonstrate its efficacy in improving the frequency response and to operate within the grid code above 59.5 Hz.

1. A simplified inertial response model of the Caribbean system (Figure 8.15) was simulated to validate the detailed UPME simulations of the system frequency response and to serve as a test bed for potential improvements.
2. The simplified simulation mimics the detailed simulations by showing the frequency drop to 59.4 Hz within three seconds.
3. The model clearly demonstrates that adding a 65-MW battery system will prevent the frequency from declining below 59.5 Hz. See Figure 8.16. The battery frequency control loop measures local bus frequency and discharges energy whenever the frequency drops and charges when it rises. The battery controller can have a droop characteristic when the frequency deviation is small, and bang-bang characteristics when the frequency deviation is high.

8.3.3 Cost-Benefit Analysis

A high-level gross economic analysis, as shown in Figure 8.17, reveals that energy storage with 30 minutes capacity supplying frequency response services

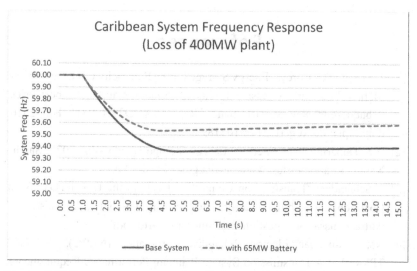

Figure 8.16 Frequency response after the loss of 400 MW and 65-MW battery.

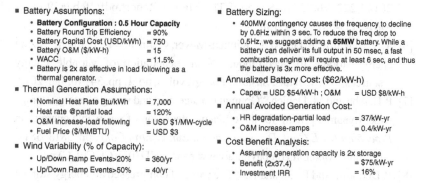

Figure 8.17 Frequency regulation economic analysis for current market conditions in Colombia

to support the wind integration can yield pretax project IRRs of 16%, given the current prevailing market prices and structure.

The economic analysis accounted for the expected savings in fuel and maintenance from utilizing a battery system instead of partially loading thermal units to follow load or counteract ramping events. The analysis did not account for the potential outage cost resulting from load shedding, which will drastically improve battery economics.

References

[1] J. Eyer and G. Corey, "Energy Storage for the Electricity Grid: Benefit and Market Potential Assessment Guide," Sandia National Laboratories, Albuquerque, New Mexico, Sandia Report SAND2010-0815, Feb. 2010. doi: https://doi.org/10.2172/1031895.

[2] U.S. Department of Energy, "Energy Storage Handbook," 2020. Sandia National Laboratories. www.sandia.gov/ess-ssl/eshb.

[3] J. Schneider, "Transmission Congestion Costs in the U.S. RTOs," Watt Coalition, 2019. https://watt-transmission.org/wp-content/uploads/2019/08/transmission-congestion-costs-in-the-u.s.-rtos.pdf.

[4] North American Electric Reliability Corporation (NERC), "Standard TPL-001-4 – Transmission System Planning Performance Requirements, Version 3," 2018. www.nerc.com/files/TPL-001-4.pdf.

[5] Potomac Economics, "2016 State of the Market Report for the New York ISO Electricity Markets," p. 9, May 2018. www.nyiso.com/documents/20142/2223763/2016-State-Of-The-Market-Report.pdf/2feb2a59-df4c-e967-0a53-6818458a3138.

[6] Y. V. Makarov, P. Du, M. C. Kintner-Meyer, C. Jin, and H. F. Illian, "Sizing energy storage to accommodate high penetration of variable energy resources," *IEEE Trans. Sustain. Energy*, vol. 3, no. 1, pp. 34–40, Jan. 2012.

[7] P. Harsha and M. Dahleh, "Optimal management and sizing of energy storage under dynamic pricing for the efficient integration of renewable energy," in *Proc. 50th IEEE Conf. Decision Control Eur. Control Conf.*, pp. 5813–5819, 2011. http://web.mit.edu/pavithra/www/papers/HD2011(CDCECC).pdf.

[8] P. Denholm and R. Sioshansi, "The value of compressed air energy storage with wind in transmission-constrained electric power sytems," *Energy Policy*, vol. 37, no. 8, pp. 3149–3158, May 2009.

[9] M. Ghofrani, A. Arabali, M. Etezadi-Amoli, and M. S. Fadali, "A frame-work for optimal placement of energy storage units within a power system with high wind penetration," *IEEE Trans. Sustain. Energy*, vol. 4, no. 2, pp. 434–442, Apr. 2013.

[10] S. Yan, Y. Zheng, and D. Hill, "Frequency constrained optimal siting and sizing of energy storage," *IEEE Access*, vol. 7, pp. 91785–91798 2019. doi: https://doi.org/10.1109/ACCESS.2019.2927024.

[11] H. Pandžić, T. Q. Dvorkin, and D. S. Kitschen, "Near-optimal method for siting and sizing of distributed storage in a transmission network," *IEEE Trans. Power Syst.*, vol. 30, no. 5, pp. 2288–2300.

[12] S. Bose, D. F. Gayme, U. Topcu, and K. M. Chandy, "Optimal placement of energy storage in the grid," in *Proc. 51st IEEE Conf. Decision Control,* pp. 5605–5612, 2012. doi: https://doi.org/10.1109/CDC.2012.6426113.

[13] S. Wogrin and D. Gayme, "Optimizing storage siting, sizing, and technology portfolios in transmission-constrained networks," *IEEE Trans. Power Syst.,* vol. 30, no. 6, pp. 3304–3313, Nov. 2015.

[14] R. Fernández-Blanco, Y. Dvorkin, B. Xu, Y. Wang, and D. S. Kirschen, "Optimal energy storage siting and sizing: a WECC case study," *IEEE Trans. Sustain. Energy,* vol. 8, no. 2, p. 733–743, Apr. 2017.

[15] H. Khani, M. R. D. Zadeh, and A. H. Hajimiragha, "Transmission congestion relief using privately owned large-scale energy storage systems in a competitive electricity market," *IEEE Trans. Power Syst.,* vol. 31, no. 2, pp. 1449–1458, Mar. 2016.

[16] S. E. Del Rosso, "Energy storage for relief of transmission congestion," *IEEE Trans. Smart Grid,* vol. 5, no. 2, pp. 1138–1146, Mar. 2014.

[17] J. Harlow and et al., "A wide range of testing results on an excellent lithium-ion cell chemistry to be used as benchmarks for new battery technologies," *J. Electrochem. Soc.,* vol. 166, no. 13, pp. A3031–A3044, 2019. doi: https://doi.org/10.1149/2.0981913jes.

[18] WECC Renewable Energy Working Task Force, "WECC Battery Storage Dynamic Modeling Guideline," Nov. 2016. https://dokumen.tips/docu ments/wecc-battery-storage-guideline-western-electricity-battery-stor age.html.

[19] O. Tremblay, L.-A. Dessaint, and A.-I. Dekkiche, "A generic battery model for the dynamic simulation of hybrid electric vehicles," in *2007 IEEE Vehicle Power and Propulsion Conference,* pp. 284–289, 2007. doi: https://doi.org/10.1109/VPPC.2007.4544139.

[20] National Grid ESO, "Electricity Ten Year Statement (ETYS)," Nov. 2018. www.nationalgrideso.com/research-publications/etys/archive.

[21] M. Joos and I. Staffell, "Short-term integration costs of variable renewable energy: wind curtailment and balancing in Britain and Germany," *Renew. Sust. Energ. Rev.,* vol. 86, pp. 45–65, 2018.

[22] G. Fitzgerald, J. Mandel, J. Morris, and H. Touati, "The Economics of Battery Energy Storage: How Multi-use, Customer-Sited Batteries Deliver the Most Services and Value to Customers and the Grid," Rocky Mountain Institute, Sept. 2015. [Online]. www.rmi.org/electricity_battery_Value.

[23] EPRI and U.S. Department of Energy, "Handbook of Energy Storage for Transmission and Distribution Applications," Final Report, Palo Alto, CA, Dec. 2003. www.sandia.gov/ess-ssl/publications/ESHB%201001834% 20reduced%20size.pdf.

[24] Renewable Energy Foundation, Blog 348, "Constraint Payments to Wind Farms." [Online]. www.ref.org.uk.

[25] National Grid ESO, "Balancing Services Use of System (BSUoS) Charges," 2018. [Online]. www.nationalgrideso.com/charging/balancing-services-use-system-bsuos-charges.

Acknowledgment

The author acknowledges the significant contributions of his colleagues at Quanta Technology in applying the various concepts during the conduct of several consulting engagements with utility customers. Special acknowledgement is reserved for Abhishek Thurumalla and Rahul Anilkumar for their distinguished contributions.

About the Author

Hisham leads the transmission and regulatory compliance consulting practice at Quanta Technology LLC, providing technical and economic advisory services supporting regulated utilities, energy developers, and RTOs to address their evolving and challenging business needs. Dr. Othman has 30 years of technical and managerial experience in the electricity sector, with an emphasis on grid integration of renewables, energy storage, and business strategy.

Hisham worked with leading teams that introduced the thyristor-controlled series capacitor, implemented the first ISO/RTO operational and business IT system in the US, implemented the largest distributed 40 MW solar system on utility poles in the world, designed and implemented a high-penetration fuel abatement solar–diesel system, and extensively modeled, analyzed, and invested in energy storage applications. Hisham holds a PhD in Electrical Engineering from the University of Illinois at Urbana-Champaign.

Cambridge Elements ≡

Grid Energy Storage

Babu Chalamala
Sandia National Laboratories

Dr. Babu Chalamala is manager of the Energy Storage Technologies and Systems Department at Sandia National Laboratories. He received his Ph.D. degree in Physics from the University of North Texas and has extensive corporate and startup experience spanning several years. He is an IEEE Fellow and chair of the IEEE Energy Storage and Stationary Battery Committee.

Vincent Sprenkle
Pacific Northwest National Laboratory

Dr. Vincent Sprenkle is chief scientist at Pacific Northwest National Laboratory (PNNL), and program manager for the Department of Energy's Electricity Energy Storage Program at PNNL. His work focusses on electrochemical energy storage technologies to enable renewable integration and improve grid support. He has a Ph.D. from the University of Missouri and holds 25 patents on fuel cells and batteries.

Imre Gyuk
US Department of Energy

Dr. Imre Gyuk is Director of the Energy Storage Research Program at DOE's Office of Electricity. For the past 2 decades, he has directed work on a wide portfolio of storage technologies for a broad spectrum of applications. He has a Ph.D. from Purdue University. His work has won prestigious awards including 12 R&D 100 Awards, the Phil Symons Award from ESA, and a Lifetime Achievement Award from NAATBatt.

Ralph D. Masiello
Quanta Technology

Dr. Ralph D. Masiello is a senior advisor at Quanta Technology, and developed smart grid roadmaps for several US independent system operators and the California Energy Commission. With a Ph.D. from MIT in electrical engineering, he is a Life Fellow of the IEEE, member of the US National Academy of Engineering, and won the 2009 IEEE Power Engineering Concordia Award.

Raymond Byrne
Sandia National Laboratories

Dr. Raymond Byrne is the manager of the Power Electronics and Energy Conversion Systems Department at Sandia National Laboratories, and works on optimal control of energy storage to maximize grid integration of renewables. He is a Fellow of the IEEE and recipient of the IEEE Millennium Medal.

About the Series

This new Elements series is perfect for practicing engineers who need to incorporate grid energy storage into their electricity infrastructure and seek comprehensive technical details about all aspects of grid energy storage. The addressed topics will span from energy storage materials to the engineering of energy storage systems. Cumulatively, the Elements series will cover energy storage technologies, distributed energy storage systems, power electronics and control systems for grid and off-grid storage, the application of stationary energy storage systems for improving grid stability and reliability, and the integration of energy storage in electricity infrastructure. This series is co-published in collaboration with the Materials Research Society.

MATERIALS RESEARCH SOCIETY®
Advancing materials. Improving the quality of life.

Cambridge Elements ≡

Grid Energy Storage

Elements in the Series